# Scrap & Build

by

D. C. E. Burrell

Published by the World Ship Society
Kendal, Cumbria LA9 7LT
ISBN 0 905617 29 0

## LIST OF CONTENTS



*(Cover)* Launch of the **NORTH BRITAIN.** *(South Tyneside Public Libraries & Museums)*

## ACKNOWLEDGEMENTS

The primary source for material on the Scrap & Build and Shipping Loan schemes are the Committee records now in the care of the Public Record Office, Kew (particularly files MT9/2766, 2767, 2770 and 3198). From this start has been added material from the Transcripts of Ownership (P.R.O. series BT110), Lloyd's Register of Shipping, Lloyd's of London Marine Collection at the Guildhall, London, and many other sources.

In compiling this record I would particularly like to thank Laurence Dunn, Paul Heaton, John Lingwood and Len Sawyer for their help, also the staff at the Public Record Office, Lloyd's Register, the Guildhall Library, Scottish Record Office, University of Glasgow, Tyne & Wear and Cleveland Archives. The Editor of "Fairplay" has kindly permitted me to quote extensively from his publication and the volumes "British Vessels Lost at Sea 1939-45" and "Die U-Boot-Erfolge, Der Achsenmachte 1939-45" have been invaluable reference guides.

So many others have assisted with advice and information including many of the shipowners and builders featured in the text that it would be difficult to prepare an exhaustive list. To all those not mentioned above, my thanks.

Photographs have come from many sources and the caption of each includes an appropriate acknowledgement.

Finally, and far from least, to Rosemary, my wife, and our family for their forebearance as this work has developed over the last few years.

Sevenoaks,
Kent. July, 1983.

## NOTES

Entries are arranged alphabetically under the name of the managers. If different the name of the owning company is shown below that of the managers.

Newbuildings:

After the name of the ship is shown the price and loan figures — it should be remembered that these are not exact but merely estimated prices used during negotiations. Unless otherwise stated the Scrap and Build loan was at 3% interest with repayment over 12 years, deferred for 2 years. The 1939 loans were at 3.75%, repayment over 12 years, deferred 6 months.

Official Number: tonnages — g = gross, n = nett, d = deadweight [these tonnages can vary over a period of time and the deadweight may be quoted differently according to what is included or excluded, i.e. bunkers and stores. The deadweights quoted are from the loan files]: Registered dimensions.

Type of engines [T 3-cyl = 3 cylinder triple expansion: C 4-cyl = 4 cylinder compound] and name of builders, power and speed as recorded in transcripts of ownership [B.H.P. and S.H.P. are practically identical and their relationship to I.H.P. is approximately 4:5]. See the appendix for information on the various types of machinery employed.

Date of launch and completion [the completion date is the month of registration]. Subsequent history and final end. Abbreviations include Nav = Navigation, Sg = Shipping, S.S. = Steamship, S.N. = Steam Navigation, Sbg = Shipbuilding, Sbkg = Shipbreaking, Tg = Trading, Hbr = Harbour, T.L. = Total Loss, C.T.L. = Constructive Total Loss, Comp.T.L. = Compromised Total Loss.

Ships scrapped:

Name, previous names, gross tonnage and year built. Name of owners (and managers), price for which purchased [note that some ships passed through a number of hands as will be apparent from the notes and entries elsewhere]. Year scrapped, by whom and price realised [the prices are as noted in 'Fairplay'].

# BRITISH SHIPPING

The rise and fall of the merchant service reflects the history of the Industrial Revolution, also that of the Empire and Commonwealth. The Napoleonic Wars so delayed industrial development that Britain was given a 50 year head start and resulted in British engineers being involved in many early industrial projects elsewhere in the world.

As the 20th century dawned the red ensign flew over more than half the ships of the world. Shipowning was an accepted investment for a large number of people, and examination of transcripts of ownership and company records reveal the wide range of people with financial interests in shipping.

During the later decades of the 19th century shipping provided a reasonable return for the investment and did not exhibit the depressions to be seen in industry. The only booms were due to the Crimea, Boer and Great Wars, but experience suggested a five year cycle, four years moderate profits followed by one year very satisfactory. In 1905 this cycle was broken largely because Burrell & Son of Glasgow, who were noted for their sagacity and timing in ordering tonnage to take advantage of good years, placed orders for 20 ships. Others followed and when trade increased in 1906 the abnormal glut of new tonnage depressed what would otherwise have been a period of good freights to produce the worst shipping depression experienced to that date. So serious was the surplus of tonnage that in 1909 a meeting of North of England shipowners at Newcastle proposed the formation of an International Union of Shipowners to regulate tonnage. The proposal envisaged all owners voluntarily laying up their ships for two months during the year, a proposal that was not adopted due to lack of support. It nevertheless gives an idea of the amount of excess tonnage seeking cargo at the time. Not until 1912 were really good freight rates obtained, followed by a relapse in 1914 which saw ships laid up prior to the outbreak of war in August.

The demands of war and the losses sustained had an artificial effect, sending freight rates rocketing. The British Government were far from generous so British owners did not benefit fully from this as they were paid the low Blue Book rates (table i), and on their profits the Government levied "excess profits duty" of up to 80 per cent of the sum by which those profits exceeded pre-war figures. On the balance left income tax was yet to be paid as well. Compensation for war losses totalled £146 million, but shipowners had to find £280 million to make good the losses. So although the Government made an estimated profit of £16 million on the war risks insurance scheme the industry returned to peace in a relatively impoverished state, so weakened financially that it suffered badly in the Depression.

**Table i**
**Examples of Blue Book rates in early 1917 compared with other flags**

| | United Kingdom | Norway | Japan Denmark | Misc. Neutrals | U.S.A. |
|---|---|---|---|---|---|
| 2,500 to 3,000dwt | 8/6 | 49/– | 45/– | 45/6 | 29/3 |
| Over 6,000dwt | 6/6 | 41/6 | 45/– | 40/9 | 24/– |

With the absurd war values some owners sold and refused to replace their tonnage until more realistic values returned. These owners, like Sir Arthur Sutherland, Sir Walter Runciman and William Tatem (The Baron Glanely) were probably the only ones really to make money out of the war. One finance house, the British Steamship Investment Trust, refused to lend finance, anticipating values would drop 50 per cent after the war. Finance houses in normal times would lend up to 50 per cent of the market value of a ship, the banks who stepped into the void lent up to 80 per cent of the abnormal values and, with little or no experience, lost heavily.

Freight rates collapsed in the summer of 1920 and some companies thereafter paid no dividend for two decades or went out of business. This especially applied to owners who had purchased tonnage at inflated wartime prices like the Greek owner(*) quoted by 'Fairplay' who, in 1919, purchased four 'B' and two 'C' class standard ships from the British Government at £289,166 and £180,000 each. Unable to operate profitably when freight rates went down he defaulted in 1922 and the vessels were auctioned. The four 'B' class realised £158,750 in total and two 'C' class £52,000, a fall in value of 86 per cent in four years.

During the 1920s the Trade Facilities Acts encouraged owners to build and keep shipbuilders employed. The arresting of the downward trend in freight rates encouraged optimism in the second half of the decade and many orders were placed. But worse was to come as the Wall Street collapse known as "Black Tuesday", 29 October 1929, heralded the Great Depression. By 1937 the world had been forced off the gold standard (Britain going in 1931). Optimistic owners paid the penalty with the giant Royal Mail Steam Packet group under Lord Kylsant heading the list, collapsing with losses in excess of £50 million. This repercussed throughout British shipping and it took years for the mess to be sorted out; for instance it was 1937 before Harland & Wolff, who were closely associated with the group, absorbed their losses by reducing their capital from £12,100,000 to a mere £1,796,082. Some shipowners with new vessels were forced to sell within four or five years at losses of 50 per cent to satisfy mortgage demands (table ii).

Statistics prepared for 'Fairplay' and 'British Shipping Finance' clearly show the picture (appendixes a and b). In general owners were living on reserves and not making provisions for depreciation. The exceptions were usually companies who had avoided buying at high prices and followed a policy of building up substantial reserves for renewals. Companies like the Nitrate Producers' S.S. Co Ltd and Tatem S.N. Co Ltd were therefore able to continue dividends earned by investments which likely subsidised the continued operation of their fleet.

**Table ii**
**Examples of ship sales — built 1929/30, sold 1934/5**

| Name | Built | Dwt | New Cost | Realised |
|---|---|---|---|---|
| PORTREGIS | '30 | 7,820 | £66,000 | £33,500 |
| KITTY TAYLOR | '29 | 8,935 | £70,000 | £42,000 |
| PORTFIELD | '29 | 7,820 | £69,000 | £35,000 |
| NAIRN | '29 | 8,764 | £77,500 | £45,000 |
| KNIGHT OF THE CROSS | '29 | 6,500 | £60,000 | £29,500 |
| STATIRA | '29 | 7,675 | £76,000 | £46,500 |

('Fairplay')

Another aspect of the problems facing world shipping in the Great Depression was the volume of world trade and the available tonnage seeking cargo. World trade in 1934 and 1935 was at about the same level as 1913, but tonnage had increased over 43 per cent. No wonder over 2 million tons were laid up on several occasions in U.K. ports alone, freights disintegrated and the statement was made that the five years to 1934 were the worst ever experienced. As 1929 was taken as the base for the shipping subsidy (although why it is difficult to understand as that year, although

**Table iii**
**Tonnage of steam and motor ships over 100 g.r.t.**

| Flag | 1914 | (in '000 tons) 1934 | % change |
|---|---|---|---|
| Great Britain and Ireland | 18,892 | 17,629 | (6.7) |
| Dominions and Colonies | 1,632 | 2,978 | 82.5 |
| | 20,524 | 20,607 | 0.4 |
| U.S.A. (excl. Great Lakes) | 2,027 | 9,795 | 383.2 |
| France | 1,922 | 3,260 | 69.6 |
| Italy | 1,430 | 2,875 | 101.1 |
| Japan | 1,708 | 4,073 | 138.5 |
| Germany | 5,135 | 3,680 | (28.3) |
| Norway | 1,957 | 3,980 | 103.4 |
| Holland | 1,472 | 2,612 | 77.4 |
| Greece | 821 | 1,507 | 83.4 |
| Others | 6,148 | 9,461 | 53.8 |
| Total | 43,144 | 61,850 | 43.4 |

(Lloyd's Register of Shipping)

**Table iv**
**Tramp Freight Index**

| Year | 1913=100 | 1920=100 | 1929=100 | | 1935=100 |
|---|---|---|---|---|---|
| 1913 | — | 23.40 | 94.09 | | |
| 1920 | 427.35 | — | 402.09 | | |
| 1929 | 106.26 | 24.87 | — | | 133.50 |
| 1930 | | | | (75.4) | |
| 1931 | | | | (78.5) | |
| 1932 | | | | (74.1) | |
| 1933 | 77.48 | 18.13 | 72.90 | (71.6) | |
| 1934 | 80.60 | 18.86 | 75.83 | (74.5) | |
| 1935 | 81.07 | 18.97 | 76.28 | (74.9) | — |
| 1936 | 96.50 | 22.58 | 90.79 | (84.3) | 112.60 |
| 1937 | 149.27 | 34.93 | 140.45 | (131.5) | 175.60 |
| 1938 | | | 95.00 | (95.0) | 126.90 |
| 1939 (8 months to August) | | | 91.20 | | 121.80 |

Note: The figures in brackets are the freight index figures used by the Board of Trade.

(Chamber of Shipping)

good for international trade, was not conspicuously profitable for shipping) we can see how world trade that year compared with the following years (1929=100):

| | | | | | |
|---|---|---|---|---|---|
| 1930 | 92.8 | 1933 | 74.9 | 1936 | 86.0 |
| 1931 | 85.3 | 1934 | 77.2 | 1937 | 98.0 |
| 1932 | 73.9 | 1935 | 79.0 | 1938 | 93.1 |

On the same basis the 1913 figure was 77.0. Table iii details Lloyd's Register figures for the increase in world shipping from 1914 to 1934.

Many countries introduced subsidies of various types, estimated by the Chamber of Shipping in 1936 as worth £30 million per annum. This had the artificial effect of keeping a world fleet in being far in excess of the needs of trade. National security was often the reason cited for this action.

The financial consequences for tramp owners can be gauged by analysis of the financial statistics prepared by 'Fairplay' and included as appendices a and b ("Cargo-Boat Earnings 1904-38" and "Workings of Some Cargo-Boat Companies in 1934"). No wonder the tramp freight index, table iv, makes abysmal reading when it is remembered that operating costs were higher than in 1920 (in 1920 operating costs on foreign voyages were estimated to be over 280 per cent higher than in 1913-14).

A number of factors influenced the rise in trade levels. Economic recovery started in 1934, became more definite the following year and shipping started to feel the effect in 1936. The improved freights peaked with a short lived boom in the summer of 1937, thereafter slumping back again until the outbreak of war in September 1939 brought an end to sanity.

As always, war provided a stimulus. Italian aspirations in Africa led to the invasion of Ethiopia in October 1935. July 1936 saw the outbreak of the Spanish Civil War and a year later the Sino-Japanese War re-erupted. The disturbed international situation caused the clouds of world war to loom again on the horizon as totalitarian forces in Europe sought their destiny, and the start of re-armament required raw materials and the ships to carry them. A poor European harvest and maize crop failure in the U.S.A. also resulted in the movement of heavier grain tonnages from South America and Australia in 1936/7.

(*)This probably referred to the Baltic Exchange auction on the 5th June 1923 of the following steamers on the instructions of The Admiralty Marshal (intended Shipping Controller names in brackets):

| | | | |
|---|---|---|---|
| AMBATIELOS (War Trooper) | £38,500 | PANAGIS (War Diadem) | £25,500 |
| STATHIS (War Piper) | £41,000 | MELLON (War Regalia) | £26,500 |
| TRIALOS (War Sceptre) | £42,500 | | |
| CEPHALONIA (War Miner) | £36,750 | | |

The four 'B' class (left hand column) and the PANAGIS had been owned by Nicolas E. Ambatielos of Argostoli, whilst the second 'C' class had been managed by the Royal Mail Steam Packet Co on behalf of the Board of Trade. The four 'B' class were all bought by King Line.

('Fairplay', 7th June 1923)

## SCRAP AND BUILD
## (The British Shipping (Assistance) Act, 1935)

Various proposals were mooted to handle the surplus of tonnage, for example in February 1933 Mr F. L. Dawson, a past president of the North of England Steamship Owners' Association, suggested that 250 British ships be purchased for £1 million and scuttled, whilst the Government made efforts at the World Monetary & Economic Conference of 1933 to obtain a measure of international agreement in achieving healthier conditions, but failed to secure support from the Governments that were subsidising their own fleets.

Finally, in November 1933, the Tramp Shipowners' Committee of the Chamber of Shipping of the United Kingdom made formal application to the Government for assistance, asking for a subsidy of 10s per gross ton per annum for tramps operating and 5s whilst laid up. The following month a Cabinet Committee was appointed to consider the problem and in May 1934 the policy of Scrap and Build was provisionally accepted.

The President of the Board of Trade stated in the Commons on the 3rd July 1934 that very few British shipping companies were covering their running expenses, and fewer still making any provision for replacement. The tramp sector was the hardest hit, hence the desire to channel assistance to tramp owners.

Various proposals were considered, the original White Paper suggesting the scrapping of three tons for each ton built, with the proviso that scrapped tonnage should have been British owned on the 31st March 1934. This proviso was not retained — it would have virtually made the Act stillborn — but was dropped as it would have caused too rapid a fall in British shipping and examination of tonnage scrapped indicates that half the vessels broken up were of foreign flag.

As the Bill developed Government concern for other areas of the economy became apparent, the final provisions including:

1) Tramp Owners — £2 million for 1935 as an operating subsidy for 'defensive purposes', with the warranty that owners formulate a scheme of co-operation and minimum freights to avoid the disastrous competition in recent years.

2) Shipbuilders — all new ships were to be products of British yards. In May 1934 over 80,000 of a workforce of 169,000 were unemployed, and production had fallen to a mere 190,476 g.t. in 1934. It was estimated that an output of 1 million tons would employ 63,000 men for a year.

3) Scrapping of old tonnage to be in the U.K. Although this was progressively relaxed it was estimated that breaking up 3 million tons would employ 15,000 men for twelve months.

The Bill drew some caustic comment from members of the Opposition in the House of Commons. Dr Addison accused shipowners of exorbitant profits, some £435 million in the first 26 months of the war, and he continued "It was not until after we had snatched the ships from the greedy hands of the owners that the nation was made more certain of being able to survive." Mr Shinwell attacked the subsidy as "providing dividends for shareholders", and other Labour members suggested a means test on owners. 'Fairplay' commented on Mr Shinwell's statement "(he) demonstrated that he had as much knowledge of shipping as a cow is presumed to have about that classical musical composition 'The Lost Chord'." Examination of the financial figures in the appendices is instructive. When the value of British shipping in time of war was cited as a reason for ensuring it was maintained at a suitable level the Lord Chancellor, Viscount Hailsham, made the naive and ludicrous suggestion that reliance should be placed on chartering neutral tonnage in time of war. The lesson of World War I seems to have been quickly forgotten. Some shipowners were also against the Act, but mainly on the grounds of Government interference in free enterprise.

To consider applications for loans in respect of modernisation or newbuildings the Ships Replacement Committee was formed. Initially Sir George Rainey was proposed for the position of chairman but the final choice was Sir Charles Innes, a retired Indian Civil Servant. The Committee had a wide range of experience as it consisted of, in addition to the chairman, a naval architect, a marine engineer, two shipowners and a businessman, whilst accountancy experience was supplied from outside the Committee by Mr G. W. C. Davis, the Board of Trade Chief Accountant. Amongst the duties incumbent on the Committee were the technical evaluation of the ships built and scrapped as well as the financial standing and eligibility of applicants for loans. Examination of the proposals for newbuildings which were declined, also the scrap tonnage offered but refused, shows that the Committee was very active in applying the conditions of the legislation during its two years existence. The Ships Replacement Committee members were (see appendix d for biographical notes):

| | |
|---|---|
| Chairman: | Sir Charles Alexander Innes, K.C.S.I., C.I.E. |
| (Naval Architect) | Frederick George Bryant, M.I.N.A. |
| (Shipowner) | George Perrin Christopher |
| (Engineer) | Albert Edward Laslett, I.S.O., M.I.N.A., M.I.Mar.E. |
| (Shipowner) | John Niven |
| (Businessman) | Major Sir Percival Reuben Reynolds, K.B.E. |
| Secretary: | A. L. Moore |

Part I of the Act provided a subsidy of £2 million for tramp voyages in 1935, provided the level of freight did not exceed 92 per cent of the 1929 level. For every 1 per cent by which the index exceeded 92 per cent the subsidy would be reduced by £250,000. This part of the Act was extended to cover 1936 and 1937 also, and was paid for the first two years. As table iv shows the index for 1937 showed a considerable recovery and no subsidy payments were therefore made. 'Fairplay' issues of the 26th March 1936 and 15th April 1937 published lists of payments in excess of £1,000.

**VIRGO,** the only "turret" ship scrapped (page 27). *(R. A. Snook)*

The payment of the subsidy was not automatic but only on recommendation of the Tramp Shipping Subsidy Committee. The only owner penalised by the Committee refusing to recommend a voyage subsidy was the Nitrate Producers' S.S. Co whose chairman, Sir John Latta, protested against the subsidy from the start although approving of minimum freights. The Committee took exception to the methods adopted by him in dealing with some of his ships contrary to the instructions they had laid down.

A major task of the associated Tramp Shipping Subsidy Committee was setting and maintaining minimum freights for a number of trades — abiding by these minimum rates was a prime requisite for subsidy payments and prevented the dissipation of the subsidy by domestic competition. Co-operation between owners in this manner could probably, if implemented in the late 1920s, have avoided the ruinous competition for starvation freights in the Depression. At the launch of the OAKDENE Mr W. Stanley Hinde, managing director of her owners, spoke of his hope that the impetus for co-operation would eliminate foolish competition and reckless individualism, but a Russian delegation in 1936 used its best endeavours to persuade owners to accept rates below the minimum for timber shipments from Russia.

Part II of the Act gave rise to the term 'Scrap & Build'. One provision was the availability of loans to modernise tonnage on the basis of scrapping one ton for every ton modernised. No owner made any proposal to the Ships Replacement Committee, probably the main reason for this was the advance in hull and machinery design that made new tonnage more desirable. Both Doxfords and Burntisland advertised and built their 'Economy' designs whilst other yards had not been idle despite lack of orders. For example Laing and Thompson had no orders for three years, but their design staff turned their attention to co-operation in experimenting and developing better designs. The first fruit of this work was the Thompson built EMBASSAGE for Hall Bros, Newcastle. Many others were built by both companies and led directly to the Liberty ship design. The EMBASSAGE had a deadweight of 9,100 tons and, in unfavourable weather, averaged 10 kts loaded on 17 tons of coal a day. A decade or so earlier consumption would have been about 25 tons. The seven ships built by these yards under the Act benefited from this co-operation in design work.

Whereas a wide variety of ship types were scrapped, single, multi and well deck, three island, long bridge and raised quarter deck, awning, shade, spar and shelter deck with one "turret", the VIRGO, the newbuildings were, with few exceptions, of modern shelter deck construction, the most economical type available for general trading. Apart from the small vessels, under about 1,500 g.t., the only exceptions to this were the single deck BIDDLESTONE and long bridge deck CRESSDENE. A number were also equipped with the latest designs of steam machinery, and more detail of some of these is given in appendix h.

As might be expected some proposals came from enterprising gentlemen who saw the opportunity for embarking on a career as shipowners with Government finance at risk. The Committee gave scant consideration to such proposals. There was also concern at the possibility of loans to companies considered unsound. One company that caused apprehension was the Silver Line which had built up a fleet at high cost under the Trade Facilities Acts. The company was experiencing stress because of the high interest payments due on its borrowing and as a result in 1936 reduced its capital from £944,970 to £23,624 (a £1 share was devalued to 6d).

Initially scrapping was only in the U.K., and early in 1935 the owners of the DOVEDEN (ex Howick Hall) sacrificed their scrap and build rights for the higher price obtainable from Italian breakers. Later the Ships Replacement Committee increasingly gave owners permission to sell to breakers abroad to obtain better prices.

Obtaining scrap tonnage was an increasing problem as the Chairman of the Burntisland Shipbuilding Co, Mr A. L. Ayre, commented at the launch of the AURETTA, two tons having to be broken up for every ton financed by a loan. In June 1936 Mr F. C. Pyman speaking at Middlesbrough highlighted the narrow interpretation that required scrapped ships to be of the same type and in the same trade as the newbuildings. Only six of the 97 ships nominated and accepted for scrapping had been the property of the firms nominating them. The market that developed for suitable tonnage pushed the prices up and effectively added 10 per cent to the cost of new ships. Early in 1935 suitable tonnage was available at 26s per g.r.t. but later in the year demand had pushed it up to 41s. On resale to breakers in the U.K. about 23/6d was realised, hence the desire to sell abroad where Italian breakers were paying 29/6d, moving up to 34/6d. However, in mid 1935 Italian breakers found difficulty in obtaining sterling and over a dozen sales were cancelled. Japan was a buyer from time to time at prices of 32/6d to 36/9d. Shipbreaking in Japan was treated as a state secret and it has therefore been difficult to identify the breakers concerned although most vessels were imported by Amakusa Sangyo Kisen Kabushiki Kaisha and broken up at Osaka by Okada Gumi, Kitagawa Gumi and Sanwa Shoji as sub-contractors.

It is interesting to see who took advantage of the Act, and who didn't. The Dunlop fleet in Glasgow was entirely renewed and others took full advantage like Stephens, Sutton Ltd, Newcastle, whose Mr R. M. Sutton spoke of the value of the scheme. Some owners who might have benefited, like Turnbull, Scott & Co and the Court Line, both of London, had already modernised their fleets and had no need for loans, but others were outspoken critics, for example Sir John Latta of Lawther, Latta & Co (managers of the Nitrate Producers' S.S. Co) and Mr E. H. Watts of Watts, Watts & Co (managers of the Britain S.S. Co). Neither of these companies built under the scheme.

## SHIPPING LOAN
## (The British Shipping (Assistance) Bill, 1939)

Following the lapse of the 1935 assistance due to the short lived revival of freight rates in 1937 the position steadily deteriorated till the tramp freight index fell below the maximum subsidy level of 1935, whilst shipbuilding costs rose by 50 per cent from 1936 to 1939. The result was that British shipowners only ordered five tramps of 24,380 g.t. in the first eleven months of 1938 and a total of three in the first three months of 1939. Projections showed that if replacement continued at that level British shipping would fall from 17.75 to 11 million tons in the next decade. At the same time during 1938 only four of Britain's 36 million tons of coal exports were carried by British flag ships and 8 per cent of timber imports.

British shipping was the poor cousin of many smaller industries when it came to grants from Government funds. Social service expenditure rose from £35.5 million in 1900 to £421.5 million in 1936. Apart from the tramp shipping subsidy of 1935-6 (£4 million) virtually the only help had been loans under the same legislation and, of course, the loan to Cunard in respect of the QUEENS. Much was made of grants in other directions, especially of over £21 million to the sugar beet industry from 1931-8, or £2.5 million per annum for an industry with a capital investment of £10 million which compared to many times that figure in shipping. Home production of beet also lost shipping an estimated £350,000 a year for freight on the replaced cane sugar.

During 1938 concern was growing in the shipbuilding industry at the drop in merchant ship orders, although re-armament was producing an increase in warship construction. The attention of the Government was drawn to the ordering, in 1937-8, of ships valued at £7 million abroad, many from German yards to utilise funds frozen in that country. Orders placed in the U.K. during 1938 were estimated at 300,000 g.t. or a third of the 1937 figure (see table v). The loss of tanker orders was of especial concern, in December 1929 tankers of 398,000 g.t. were building in U.K. yards and 230,000 g.t. abroad whereas in September 1939 the figures were 229,000 g.t. and 598,000 g.t. respectively.

**Table v**
**Orders placed with U.K. shipyards**

| Year (ending December) | Number of Vessels | Gross Tonnage | 1929=100 |
|---|---|---|---|
| 1929 | 408 | 1,384,129 | — |
| 1930 | 192 | 597,412 | 43 |
| 1931 | 87 | 214,366 | 15 |
| 1932 | 55 | 102,342 | 7 |
| 1933 | 125 | 315,066 | 23 |
| 1934 | 209 | 491,262 | 35 |
| 1935 | 302 | 731,642 | 53 |
| 1936 | 338 | 1,091,446 | 79 |
| 1937 | 243 | 854,095 | 62 |
| 1938 | estimated | 300,000 | 22 |

As proposals developed they encompassed a wider field than the previous legislation, including aid to threatened liner services. Government lethargy was still only too obvious in this field — still fresh in mind was the case of the Imperial Shipping Committee recommendation of aid to the Union Steamship Co of New Zealand to build two new ships to compete against U.S. subsidised services on the route from North America to the Antipodes. Government procrastination was such that when a decision was made to give assistance building costs had risen, no orders resulted and the long established service closed.

On the 28th March 1939 Mr Oliver Stanley, President of the Board of Trade, outlined his proposals. It was anticipated that some time would elapse before owners accepted the scheme, but applications flooded in, 37 cargo liners (210,000 g.t.) and 110 tramps (495,000 g.t.) in three weeks. The Board of Trade was unprepared for the response and announced that no further notifications of orders for financing under the scheme would be accepted. 'Fairplay' commented "the scheme appears . . . to have been grossly mismanaged . . ."!

The Bill developed to provide:

1) Tramp Owners — an operating subsidy of £2.75 million p.a. for five years, 1940 to 1944. The conditions were similar to the 1935 subsidy with payments decreasing between 95 per cent and 105 per cent of 1929 freight levels. One important change was the inclusion of cargo liners and short sea tramps.

2) Shipbuilding Loans — £10 million available for vessels ordered on or after the 29th March 1939 on similar terms to 'Scrap and Build', except that the scrapping clause was deleted. The loans also extended to the wider range of ships mentioned above.

3) Shipbuilding Grants — £2.5 million spread over five years, also in part linked to the level of the freight index.

4) Assistance to the Liner Services — £10 million available to support services suffering from subsidised foreign competition.

5) Merchant Shipping Reserve — £2 million to purchase British flag ships and maintain them in reserve for emergency use.

The reaction was mixed, especially as the extent of orders became apparent and the effect this would have in diluting the funds for shipbuilding grants was realised. Many owners had expected the Government to meet the difference in current newbuilding costs compared with 1936. One estimate was that the cost of newbuilding had risen by 25 per cent from January 1937 to December 1938, whilst another gave the price per ton deadweight for tramps as being £9 in 1935 and £14 in 1938. Hopes were dashed and a number of orders were cancelled when the level of aid became known.

Again there was much debate in Parliament, the opposition Labour Party levelling the same charges of profiteering against shipowners. In November 1939, during the debate on the new wartime Ministry of Shipping, not on the Shipping Assistance Bill, Mr Shinwell surprised everyone with the statement "Either the Government must give to shipowners a reasonable profit or they must take over the whole of our shipping . . ." and later "It is true that many tramp shipping firms are incurring losses at this time . . .". Comments like this and "We have to be fair, even to shipowners" brought horror to, and denunciation from, fellow Party members.

The Bill had reached the second reading when the outbreak of war caused abandonment. The immediate adoption of State control by requisitioning shipping indicated that all the lessons of World War I had not been forgotten. The free shipping market disappeared overnight and was replaced by new Blue Book rates. With it went the need for the tramp shipping subsidy and liner defence scheme, also the merchant shipping reserve.

The Reserve had already started with the purchase of four vessels, renamed with the prefix BOT- (compiled from the initials of the *B*oard *o*f *T*rade). A total of twelve had been recommended for purchase, of which two were withdrawn by their vendors leaving ten costing £225,500. The ships under negotiation were taken over on behalf of the Admiralty and later managed by their old owners on behalf of the Ministry of Shipping. These vessels are detailed in appendix m.

Much thought was given to the other two provisions of the Bill, the shipping loan and grant. Orders had been placed by owners following their understanding with the Board of Trade earlier in the year and the Government considered that to declare them obsolete due to the outbreak of war would be a breach of confidence. On the 3rd October the announcement was therefore made that loan applications would be considered for orders placed between the 29th March and 3rd October (both dates inclusive). Initial applications from 47 owners for 69 vessels were received.

The Shipbuilding Loans Committee, of Treasury and Ministry of Shipping representatives, was established to handle the applications. The Committee was not given the breadth of technical experience of the Scrap and Build Committee as technical specifications were being vetted under wartime regulations and the Committee's responsibility was largely limited to the financial standing of applicants. The initial choice as chairman was Sir David Owen, Lord Catto and Sir Charles Innes also being considered, before the Committee was finally chosen, viz

| | |
|---|---|
| Chairman: | Sir Alan Rae Smith, O.B.E.<br>(Financial Adviser to the Ministry of Shipping) |
| Ministry of Shipping: | Frank Charlton (Furness, Withy & Co)<br>George Perrin Christopher (Hain S.S. Co)<br>George Chester Duggan, C.B., O.B.E.<br>Sir John Niven, Kt (Andrew Weir & Co) |
| Treasury: | Philip Dennis Proctor |
| Secretary: | C. T. Plumb |

**ENGLISH TRADER** (ex Arctees), 3,953/34, illustrates the "arcform" convex hull sides after grounding off Dartmouth on 23.1.1937 and being refloated a month later with the loss of her bows. *(L. Dunn)*

Of the applications several were ineligible as the vessels were built in Northern Ireland which, as it still had its own Loan Guarantee operating, was excluded from the Scheme. An enquiry was even received from the Westralian Farmers Co-Operative who usually chartered tonnage to carry their exports but wished to build two ships. Although backed by the colonial government the suggestion was rejected out-of-hand as the legislation applied solely to Great Britain. Unless otherwise stated in the fleet list the refusal of loans was because the financial position of the applicants afforded inadequate security. A number of loans were made conditional, for example the capital was to be increased or the loan was to be made to the parent or associate company.

As under the earlier scheme tramps of about 5,000 g.t. featured largely in the list of buildings, although a number of over 7,000 g.t. will be found. These were not, in fact, larger vessels but a result of the wartime policy of closing shelterdecks for operational convenience in the prevailing wartime emergency.

Technically the vessels built exhibited more standardisation than those built under the Scrap and Build scheme, especially in respect of machinery. The 22 Scrap and Build steamers over 1,500 g.t. included ordinary triple expansion (9), turbine (1), compound (9), turbo-compressor (5) and reheater (2) engines, compared with only simple triple expansion (22) and reheater (8) machinery in to 30 similar sized ships built with finance under the Shipping Loan scheme. Doxford opposed piston diesels powered 28 of the larger motorships, the only non-Doxford vessel under either scheme being the HYLTON which had a NEMEC diesel. The other interesting feature was the hull design of four ships for Counties Ship Management which adopted Isherwood's 'Arcform'.

## CONCLUSIONS

The Scrap and Build scheme was not an unqualified success: it was not really popular amongst shipowners as the cost premium attaching to scrap tonnage deterred many. It failed to reduce competition due to foreign involvement, although it did result in much continued co-operation and the setting of minimum freight levels. Some of the ships scrapped were no longer capable of going to sea and were due for breaking up in any case, whilst at the end of the scheme lack of suitable tonnage was apparent.

The addition of 50 modern ships was undoubtedly a good thing, but sadly many were not to have a long life as the holocaust of war again swept over the world in September 1939. Before peace returned in 1945 over half of the ships were to fall victim to enemy action.

The Shipping Loan and Grant scheme never had an opportunity to show results as the outbreak of war brought it to a premature end. However, the number of orders placed and loans made in the few months involved serves to show how the scrapping provision stunted the acceptance of Scrap and Build. Like the earlier schemes the war brought over half of the ships to a premature end, as detailed in table vi.

**Table vi**
**Ultimate end of ships built**

| | Scrap and Build | | Shipping Loan | |
|---|---|---|---|---|
| War loss | 27* | 118,276gt | 25** | 117,526gt |
| Broken up | 15 | 50,542gt | 18 | 84,400gt |
| Wrecked/Stranded | 4* | 10,236gt | 2 | 10,356gt |
| Abandoned/Beached | | nil | 1 | 567gt |
| Missing | 1 | 5,267gt | | nil |
| Burnt | 1 | 4,993gt | | nil |
| Foundered | 2 | 2,113gt | 1** | 868gt |
| Collision | | nil | 1 | 1,911gt |
| Still afloat | 1 | 538gt | | nil |

Ships dismantled following damage have been included in the broken up entry.
*The USKSIDE is included in both figures. She was a total loss by enemy action, salvaged and rebuilt, and finally wrecked.
**The WORTHTOWN is included in both figures, being a war loss, salved and finally foundering.

Whilst it is difficult to quantify the value of the ships in the 1939-45 War certain conclusions can be drawn. Of the ships scrapped 49 of 238,000 g.t. were British, ranging from the LANCASHIRE of 1892 to the INDIAN CITY of 1920. In 1935 their average age was over 24 years. The value of many of these vessels would have been open to question especially under the stress of war if they had not previously been scrapped or sold abroad.

The Scrap & Build Scheme saw them replaced by 50 new, more reliable and economical ships of 189,000 g.t. Would they have been built without Government encouragement? For many the answer must be negative as the owners would probably have had difficulty in raising finance. Although repayment problems were only experienced with one owner the rating of the risk attached to the loans may be gauged when the 2.75 to 3% interest charge is compared to the Bank Rate which remained at 2% from June 1932 until doubling in August 1939. The Government's willingness to finance the loans, supported by the Tramp Shipping Subsidy, provided a backing which would probably not have been available commercially in view of the depressed freight market and the consequent impairment of many owners' ability to service their debts.

The disruption of shipping by the war and loss of capacity are discussed in "Merchant Shipping and War" by M. Doughty. Diversion of shipping, convoy delays and port congestion were all responsible, along with the more obvious demand for extra raw materials and military service as well as well as losses from enemy action, offset to a small degree by the suppression of luxury trades. Under such circumstances these modern vessels must have been invaluable.

The construction of the Shipping Loan vessels was, in many cases, well advanced on the outbreak of war and half were in service by the summer of 1940. Some time was to elapse before Government orders could be filled and it was to be 1941 before many British built Empire ships were operational, followed in 1942 by the Oceans and Forts ordered from the United States and Canada.

The Depression, however, did produce the Schierwater Plan conceived and promoted in 1934 by independent tanker owners and aimed at reducing surplus tonnage under the administration of Intertanko (The International Tanker Owners' Association). Owners securing charters paid a proportion of their freight earnings into a fund used to recompense owners who laid up their vessels. Although not members the major oil companies gave their support to the Plan by giving preference to vessels entered in it, hence ensuring success. Within a few years the improved tanker market led to the Plan being suspended whilst the War completely altered conditions.

Following World War II came a decade of good freights aided by the Korean War in 1951 and closure of the Suez Canal in 1956. The 1960s saw container, ro-ro, bulk and other specialist shipping eroding the place of the general cargo and tramp ship. The Suez Canal again closed in 1967, this time not to re-open until 1975. O.P.E.C. became a force to be reckoned with and assumed control of oil pricing policies, whilst Arab oil producers commenced using oil in international politics with their embargo on the U.S.A. following the 1973 war between Egypt and Israel. The quadrupling of oil prices in 1973 precipitated the tanker slump and world recession, inevitably affecting the dry bulk and liner trades.

Various schemes aimed at reducing excess tonnage have been aired since, but have simply not been adopted or not worked. As a result of the tanker market collapse following the Suez Canal re-opening in 1957 the International Tanker Recovery Scheme was promoted in 1963 to compensate owners laying up ships and encourage the scrapping of obsolete tonnage. But unlike the Schierwater Plan it failed, mainly because of lack of support by the major oil companies. Attempts to gain co-operation and control of excess tonnage continued but foundered partly on U.S. anti-trust laws.

Winston Churchill called the United Nations "the brawling cockpit of ideologies" and shipowners exhibit no different characteristics from their politicians. If anything the dispersal of shipowning amongst an increasing number of countries, aided by the policies of U.N.C.T.A.D. and worldwide depression, has made agreement less likely. There will always be an owner with lower costs, or at the instigation of politicians, who will see the opportunity to step in when attempts are made to avoid the ill effects of excess tonnage. Probably only supply and demand along with financial strength or political protection will decide in the long term.

**MARIA PRECA,** sole survivor of the ships built under the two Schemes, was originally the CROMARTY FIRTH (page 31). Note her superstructure rebuilt after sale to Greek owners.

*(W. Cassar)*

## DOXFORD'S "ECONOMY" SHIPS

William Doxford & Sons Ltd., Sunderland, were the builders who obtained the largest number of orders under the Scrap & Build and Shipping Loan schemes.

Better known for their earlier "turret" ships — see the VIRGO on pages 9 and 26 — and opposed piston diesels, Doxford's design department produced their up-to-date "economy" design in 1934. She was a 9,000 ton deadweight vessel powered by a 1,800 b.h.p. Doxford diesel with a daily oil consumption of 6.5 tons at 10 kts. Ten of the 23 built were financed under the schemes as also were six of the improved 1938 design which had a deadweight of 9,500 tons and a 2,500 b.h.p. Doxford diesel giving 12 kts on 9.5 tons of oil. The most obvious difference was the split superstructure of the 1938 design, of which 85 were built up to 1954, compared with the compact layout of the earlier group.

(Above) The RILEY (page 39) illustrates the 1934 design and (below) the DUKE OF ATHENS (page 52) typifies the later 1938 vessels. *(W. H. Brown and Skyfotos).*

Sunderland, once known as "The largest shipbuilding town in the world", was also responsible for the DORINGTON COURT, prototype of the Liberty ships, and the SD14 Liberty ship replacement.

# SCRAP AND BUILD — FLEET LISTS

**Ambrose, Davies & Matthews Ltd, Swansea**
**Cook Sg Co Ltd**
Proposed a Burntisland built vessel of 5,500g, 9,500d, costing £75,000.
No recommendation made.
Two vessels were purchased and offered for scrapping, viz
*GIEKERK* (ex Bawean), 6,427/14. Vereenigde Nederlandsche Scheepv Maats (Directie- en Agentuur Maats "Holland-Britisch-Indie Lijn" N.V.), The Hague.
*BRYNMEL* (ex Bradburn, Atlantic City), 4,707/12. Anglo-Celtic Sg Co Ltd (Griffiths, Payne & Co Ltd), Cardiff.
Notes: The BRYNMEL, purchased for £6,250, was sold for £7,650 to Greek owners for further trading as the MARIA L.

GIEKERK. *(World Ship Photo Library)*

The GIEKERK, purchased for £9,375, was resold to A. Lauro, Naples, for scrapping, the sale price being £10,000. However, she was renamed LIANA and kept trading, resulting in a dispute as reported in 'Fairplay' (16th July 1936):
"Events arising out of a purchase under the 'scrap-and-build' scheme have interesting features. A British owner, in order to obtain the 'two tons for one', purchased a Dutch steamer, giving an undertaking to break up and not trade the vessel, but found, when she was delivered, that she was in a much better condition than he had anticipated. He at once approached the owner and asked him how much he would take to delete the breaking-up clause in the contract. The Dutch seller, however, asked so much that the British owner refused to consider the offer, and sold the ship to an Italian on the same conditions of contract as he had acquired her. When the vessel arrived at the Italian port, the new buyer, like the previous one, realised that she was in a much better condition than he had expected, and reckoned that there would be profit in running her for several voyages. Accordingly he approached the seller, and, finding no encouragement in that quarter, applied direct to the late Dutch owner, offering £400 to cancel the undertaking to break up, the vendors offering to take £1,000. Considering the terms demanded too onerous, the buyer decided to run the vessel without regard to the terms of his contract. Procedure along that line naturally did not suit the Dutch owner. He proceeded to sue the British buyer, who suggested joint action against the Italian, which the Dutch owner declined. In the end, the matter was referred to a well-known arbitrator in London, the Italian buyer tendering £50 in satisfaction of the claim. The arbitrator gave his award in the form of a 'special case,' in which he found that the sellers had suffered only 1s damages, and ordered them to repay the owners the £50 which had been tendered in satisfaction of the claim. The British owner thereupon again suggested joint action against the Italian, which was then agreed to, and a threat to bring the matter before the committee of the Baltic Exchange was issued, as the agents of the Italian owner were members of that institution. This impressed the Italian owner sufficiently to prompt him to offer at once payment in settlement against the breaking-up clause being waived. The matter has now been settled by a cash payment from the Italian to the British owner, which he has shared equally with the Dutch owner."

**Anning Brothers, Cardiff**
**Exmouth S.S. Co Ltd**
**STARCROSS** Price/Loan: £71,000/£47,500
O.N. 162115. 4,693g. 2,748n. 9,000d. 411.0 × 57.7 × 24.4 feet.
T 3-cyl by Richardsons, Westgarth & Co Ltd, Hartlepool. 1,250 I.H.P. — 10 kts.
17.9.1936 launched and 11.1936 completed by Joseph L. Thompson & Sons Ltd, Sunderland.

**STARCROSS.** *(E. N. Taylor)*

20.5.1941: torpedoed by Italian submarine OTARIA, convoy SL73, in 51.45N 20.45W. Derelict sunk by escort. Lagos & Freetown for Hull, butter, groundnuts & seeds. 40 crew, all saved.
Scrapped:
*HAVILDAR,* 4,911/11. Asiatic S.N. Co Ltd, London (£9,500). 1936 — S. Okada, Osaka (£8,000).
*WEARBRIDGE,* 4,431/11. North of England S.S. Co Ltd (Crosby, Son & Co Ltd), West Hartlepool (£6,800). 1937 — Metal Industries Ltd, Rosyth (£6,250). Demolition began 30.6.1937.

**B. & S. Shipping Co Ltd, Cardiff**
**St Quentin Sg Co Ltd**
**ST HELENA** Price/Loan: £83,500/£83,500
O.N. 162140. 4,313g. 2,605n. 8,000d. 399.0 × 56.0 × 22.3 feet.
White C 4-cyl plus LP turbine by builders. 1,650 I.H.P. — 10 kts.
8.4.1936 launched and 7.1936 completed by Joseph L. Thompson & Sons Ltd, Sunderland.
12.4.1941: torpedoed by U124 in 7.50N 14W. Montevideo & Bahia for Hull, grain & general. 38 crew, all saved.
Scrapped:
*CHARLBURY* (ex Sakkarah), 4,691/06. Alexander Sg Co Ltd (Capper, Alexander & Co), London (£7,400). 1936 — Osaka (£10,500).
*MARIA,* 3,090/01. C.E. Lemos, Chios (£5,100). 1936 — Metal Industries Ltd, Rosyth (£4,500). Demolition began 18.3.1936.

**ST MARGARET** Price/Loan: £83,000/£83,000
O.N. 162141. Details as ST HELENA except 4,312g. 2,604n.
21.5.1936 launched and 8.1936 completed by Joseph L. Thompson & Sons Ltd, Sunderland.
27.2.1943: torpedoed by U66 in 27.38N 43.23W. Liverpool for Pernambuco & Buenos Aires, general & coal. 43 crew & 7 passengers, 3 crew lost and master p.o.w.
Scrapped:
*COALBY* (ex Kamouraska), 4,904/11. Ropner Sg Co Ltd (Sir R. Ropner & Co Ltd), West Hartlepool (£10,000). 1938 — Osaka (£9,750).
*GANTOISE* (ex Ripley), 3,939/02. J.A. Zachariassen & Co, Helsinki (£7,250). 1936 — Hughes Bolckow Sbkg Co Ltd, Blyth. Arrived 28.12.1935 (£6,100).

**Barry Sg Co Ltd**
**ST CLEARS** Price/Loan: £83,000/£83,000
O.N. 162142. Details as ST MARGARET.
7.7.1936 launched and 9.1936 completed by Joseph L. Thompson & Sons Ltd, Sunderland.
1951: Franz L. Nimtz, Hamburg, renamed KONSUL NIMTZ. 1960: Pan Norse S.S. Co Ltd, Panama (Wallem & Co Ltd, Hong Kong), renamed NORELG. 1.9.1962: ashore during typhoon "Wanda" on E coast of Ma Sze Chau Isl., Tolo hbr, Hong Kong. C.T.L., sold to Patt Manfield & Co Ltd and breaking up commenced 18.12.1962.
Scrapped:
*NERBUDDA,* 7,963/19. British India S.N. Co Ltd, London (£17,500). 1936 — Societa Italiana Ernesto Breda, Venice (£15,150).
*RODSKAR* (ex Rodskjaer, England, Bentala), 2,767/89 (part). Rederi A/B Rodskar (G.B. Thorden), Helsinki (£4,750). 1936 — Joseph John King & Co, Gateshead (£3,800).

**ST ROSARIO** Price/Loan: £90,000/£90,000
O.N. 162145. Details as ST MARGARET.
22.9.1937 launched and 12.1937 completed by Joseph L. Thompson & Sons Ltd,Sunderland.

**ST. CLEARS.** *(F. Barr)*

20 and 22.2.1941: bombed by Condor of KG40, convoy OB287, in 58.50N 11.40W and 59.40N 12.40W. Put back to Rothesay Bay, sailed again from Clyde 9.3.1941. Hull for Rio de Janeiro, coal. 1952: Otto Bancks Rederi A/B, Helsinki, renamed KATIA. 1958: re-engined, oil engine by Nydqvist & Holm A/B, Trollhattan, 2,000 B.H.P. — 12 kts, renamed KATIA BANCK. 1965: Ypermachos Cia Nav S.A., Panama (G. Lemos Bros. Co Ltd, London) renamed YPERMACHOS. 1968: (Transatlantic Seaways Ltd, London). 3.8.1969: arrived Whampoa for breaking up.

Scrapped:

*ST WOOLOS* (ex City of Bristol), 6,741/12. Hall Line Ltd, Liverpool (Ellerman Lines Ltd, London) (£26,000). 1938 — A. Kitagawa & Co, Osaka (£20,000).

*SYRIE* (ex Clara Zelck), 2,100/04. Cie Nat Belge de Transports Maritimes (Armement Deppe S.A.), Antwerp (£9,500). 1937 — Thos W. Ward Ltd, Barrow. Arrived 21.10.1937 (£5,650).

Notes: Management of the Barry Sg Co was transferred from Lewis Hall & Co Ltd to B & S Sg Co Ltd in 1936. Barry Sg was later renamed the South American Saint Line.

The intended name for the ST CLEARS was the ST CURIG.

The RODSKAR was purchased in 1935 and resold to E.R. Management Co Ltd (which see) for £5,500, being shared between both companies scrap and build programmes.

**CHARLBURY.** *(G. E. P. Brownell)*

**British Isles Coasters Ltd, Cardigan**

Proposed a 250g, 300d motor coaster costing £14,000. Not proceeded with.

Suggested for scrapping their CASTLEGREEN (ex Toivo, Hoffnung, Niels, Cabo Corrientes, West Coast, River Tay) (510/02). She was sold to Estonia for further trading in 1936 as the KAIDA.

**T & J Brocklebank Ltd, Liverpool**

As they proposed scrapping the MEDIA (5,437/11) they asked if she would be allowed against future building. The MEDIA was sold to the Heston Sg Co Ltd (which see).

**Wm Brown, Atkinson & Co Ltd, Hull**

Proposed a 4,970g, 9,200d motor tramp from Doxford costing £110,000, and a 1,500g, 2,500d motor tramp costing £37,000.

Proposals abandoned owing to inability to obtain scrap tonnage by 25,2,1937.

Notes: These were probably the SKIPSEA (4,974/36) and PORTSEA (1,583/38) owned by the Sea S.S. Co Ltd.

**Burnett S.S. Co Ltd, Newcastle**
Proposed two 3,100g, 4,750d steamers from Laings at a cost of £61,000 each.
Proposals abandoned owing to inability to obtain scrap tonnage by 25.2.1937.
Notes: The first application was discarded and the second was the WALLSEND (3,157/37).

**John H. Campbell**
Proposed an 800g, 1,200d tramp costing £26,000.
No recommendation.

**John Carter (Poole) Ltd, Poole**
**PURBECK** Price/Loan: £6,475/£5,500 (deferred 1 yr)
O.N. 164874. 210g 106n. 240d. 99.7 x 24.0 x 8.1 feet.
Oil engine (aft) by Atlas Diesel A/B, Stockholm. 180 B.H.P. — 7.5 kts.
19.5.1936 launched and 6.1936 completed by Charles Hill & Sons Ltd, Bristol. 1950: Lockett Wilson Ltd, London. 1959: Cerebos Ltd, London. 1961: A. J. Brush Ltd, Maldon. 1970: broken up.
Scrapped.
*LANCASHIRE,* 437/92. T. G. Irving Ltd, Sunderland. 1936 — Thos Young, Sunderland.

PURBECK. *(John Carter (Poole) Ltd.)*

**James Chambers & Co Ltd, Liverpool**
**Lancashire Sg Co Ltd**
**LOWTHER CASTLE** Price/Loan: £88,000/£75,000
O.N. 164311. 5,171g. 3.028n. 9,250d. 437.4 x 57.1 x 24.3 feet.
T 3-cyl 'Reheater' by North Eastern Marine Engineering Co Ltd, Newcastle. 1,800 I.H.P —11 kts.
12.12.1936 launched and 2.1937 completed by Sir James Laing & Sons Ltd. Sunderland. 27.5.1942: torpedoed at 5.50pm by aircraft approx. 60 miles ESE of Bear Island, convoy PQ16, caught fire and blew up 3am 28.5.1942. Sunderland and Reykjavik for Murmansk & Archangel, military stores. 54 crew, master lost.
Scrapped.
*PETRAKIS NOMIKOS* (ex Michael N, Aegaeon, Hagen), 4,418/06. P.M. Nomikos, Piraeus (£9,200). 1936 — Hughes Bolckow Sbkg Co Ltd, Blyth. Arrived 15.8.1936 (£7,650).
*SPILSBY,* 3,673/10. Ropner Sg Co Ltd (Sir R. Ropner & Co Ltd), West Hartlepool. (£7,200). 1936 — Hughes Bolckow Sbkg Co Ltd, Blyth. Arrived 22.5.1936 (£5,500).
*SAINT ANDRE,* 5,343/12 (part). Cie Generale Transatlantique, Paris (£11,000). 1937 — Ditta Luigi Pittaluga, Genoa (£8,875).

**LANCASTER CASTLE** Price/Loan: £88,000/£75,000
O.N. 164318. Details as LOWTHER CASTLE except 5,172g, 3,029n.
11.2.1937 launched and 3.1937 completed by Sir James Laing & Sons Ltd, Sunderland. 24.3.1942: bombed at Murmansk, engine room fire, extinguished. 57 crew, 10 killed. 14.4.1942: bombed and sunk in river at Murmansk. 9 killed. Master lost on passage home in H.M. ship.
Scrapped:.
*SAINT ANDRE* (part) — see above.
*LONDON CITIZEN* (ex Valemore), 5,388/18. Furness, Withy & Co Ltd, London (£12,100). 1936 — Metal Industries Ltd, Rosyth (£9,250). Demolition began 14.10.1936.

LANCASTER CASTLE. *(J. McRoberts)*

**Charter Sg Co Ltd, Cardiff**
Proposed a 4,772g, 9,000d vessel from Thompson, costing £75,000. No recommendation.

**Joseph Constantine Steamship Line Ltd, Middlesbrough**
**Constantine Sg Co Ltd**
**WINDSORWOOD** Price/Loan: £101,469.7.9d/£101,469.7.9d (interest at 2.75 per cent)
O.N. 164828. 5,395g, 3,146n. 8,550d. 438.9 x 55.8 x 24.8 feet.
T 3-cyl plus Gotaverken turbo-charger by North Eastern Marine Engineering Co Ltd, Newcastle. 1,550 I.H.P. — 10 kts.
22.4.1936 launched and 6.1936 completed by Hawthorn, Leslie & Co Ltd, Newcastle. 25.6.1940: torpedoed by U51 in 48.31N 14.50W. Tyne for Sierra Leone, coal. 40 crew, all saved.
Scrapped:
*TWICKENHAM,* 4,891/12. Britain S.S. Co Ltd (Watts, Watts & Co Ltd), London (£7,250). 1936 — Ditta Luigi Pittaluga, Genoa (£7,100).
*SURAT* (ex Betwa), 3,819/17. Bank Line Ltd (A. Weir & Co), London (£5,650). 1936 — Ditta Luigi Pittaluga, Genoa (£6,500).
*TEESPOOL,* 4,577/05 (part). Pool Sg Co Ltd (Sir R. Ropner & Co Ltd), West Hartlepool (£6,000). 1936 — Metal Industries Ltd, Rosyth (£6,400). Demolition began 1.4.1936.

SURAT. *(H. E. Weeden)*

**YORKWOOD** Price/Loan: £101,800/£101,800 (interest at 2.75 per cent)
O.N. 164829. Details as WINDSORWOOD except 5,401g. 3,150n.
21.5.1936 launched and 7.1936 completed by Hawthorn, Leslie & Co Ltd, Newcastle. 1941: Joseph Constantine S.S. Line Ltd, Middlesbrough. 8.1.1943: torpedoed by U507 in 4.10S 35.30W. Durban & Table Bay for Paranam and U.K., ballast. 48 crew, 1 lost and master p.o.w.
Scrapped:
*TEESPOOL* (part) — see above.
*ANTONIOS D KYDONIEFS* (ex Turckheim, Baycross, Crossby), 4,256/07. D. A. Kydoniefs, Andros (£6,000). 1935 — Thos W. Ward Ltd, Jarrow. Arrived 19.7.1935 (£4,400).

**YORKWOOD.** *(E. N. Taylor)*

*ANASTASIOS PETROUTSIS* (ex Despina, Raithwaite), 3,025/99. C. A. Petroutsis, Piraeus (£4,700). 1935 — Metal Industries Ltd, Charlestown (£4,350). Demolition began 26.6.1935.

**BALMORALWOOD** Price/Loan: £120,359/£120,359.
O.N. 164835. 5,832g, 3,373n. 9,200d. 446.7 x 56.7 x 26.1 feet.
T 3-cyl plus Götaverken turbo-charger by North Eastern Marine Engineering Co Ltd, Newcastle. 1,650 I.H.P. — 10 kts.
12.12.1936 launched and 2.1937 completed by Hawthorn, Leslie & Co Ltd, Newcastle. 14.6.1940: torpedoed by U47 in 50.19N 10.28W, straggler from convoy HX48. Abandoned and sunk. Sorel for Falmouth, wheat and 4 aircraft. 47 crew all saved.
Scrapped:
*AEOLOS* (ex Panagos Lyras, Panagos, Marie Z Michalinos), 3,051/02. M. N. Lyras and J. Kurmelis, Piraeus (£6,850). 1936 — Metal Industries Ltd, Rosyth (£4,750). Demolition began 25.3.1936.
*ATLANTEN* (ex Gribskov, Albuera), 3,492/02. Rederi A/B Norra Atlanten, Abo (£7,500). 1936 — John Cashmore Ltd, Newport (£5,200).
*GEORGE J GOULANDRIS* (ex Berlin, Regina), 2,187/96. N & G Simbouras, Piraeus (£4,350). 1936 — N.V. Holland Scheeps en Maschinenhandel, Hendrik-Ido-Ambacht (£3,100).
*K. LYRAS* (ex Merchant Prince), 3,091/02. P. Lyras, Oinoussai (£6,750). 1936 — Van Heyghen Freres, Ghent (£4,750).
Notes: The TEESPOOL was reported sold to Spanish breakers in the autumn of 1935 at a price of £6,500, but this came to nothing.
The TWICKENHAM, ANTONIOS D KYDONIEFS and ANASTASIOS PETROUTSIS were reported as being sold to Branch Nominees Ltd (managed by E. Atkinson), London. Details of their involvement in and relationship with Constantine are not known.

**CRESSDENE.** *(R. M. Parsons)*

**Dene Shipmanagement Co Ltd, London**
**Cressdene Sg Co Ltd**
**CRESSDENE** Price/Loan: £61,100/£61,000 (interest at 2.75 per cent)
O.N. 164608. 4,270g 2,613n. 7,555d. 385.5 x 53.2 x 24.6 feet.
T 3-cyl by Central Marine Engine Works (Wm Gray & Co Ltd), West Hartlepool. 1,500 I.H.P. — 10 kts.
24.2.1936 launched and 4.1936 completed by Wm Gray & Co Ltd, West Hartlepool. 24.5.1941:

bombed and machine gunned by aircraft in Mumbles Roads. 16.3.1942: mined 52.8N 1.52E. London for Methil & Rosario, ballast. 40 crew, all saved. Taken in tow, sank 17.3.1942 82° 2.3 miles from Sunk Lightvessel. Wreck dispersed.
Scrapped:
*NUBIAN,* 6,384/12. Fredk Leyland & Co. Ltd., Liverpool (£10,750). 1936 - Soc Italiana Ernesto Breda, Marghera (£10,750).
*EVERELSA* (ex Kamfjord, Achecolanda), 2,210/99. F. Grauds Sg Co Ltd, Riga (£3,000). 1936 — South Stockton Sbkg Co, Stockton (£2,600).
Proposed a second vessel, a 4,290g. 7,555d steam tramp from Gray, at £61,100. Application not proceeded with.
Notes: This second vessel was probably the TORDENE (4,271/36), OAKDENE (4,255/36) or FELLDENE (4,260/37).
In 1937 the Dene Sg Co Ltd was formed to take over the various single ship companies and the management firm, including the above mentioned.
Walter S. Hinde, a director, also made application under his own name (which see).

**NUBIAN.** *(World Ship Photo Library)*

**Donaldson Brothers & Black Ltd, Glasgow**
**Donaldson Line Ltd**
Enquired if they would be allowed credit on two steamers being scrapped, the NORTONIAN (6,367/13) and NUBIAN (6,384/12). These two ships had been bought in 1934 from Frederick Leyland & Co Ltd.
The proposed newbuilding was a 6,000g. 7,800d. liner. Although the matter did not proceed further the new vessel may have been the SALACIA (5,495/37).
Of the vessels proposed for scrapping the NORTONIAN went to Italian breakers in 1935 for £10,550, whilst the NUBIAN was sold to Cressdene Sg and utilised in their scrap and build application (which see).

**Douglas & Ramsey, Glasgow**
**Carrick Sg Co Ltd**
**DARLENY** Price/Loan: £89,300/£89,300
O.N. 164107. 5,205g. 2,126n. 9,400d. 432.0 x 56.2 x 25.6 feet.
T 3-cyl plus Gotaverken turbo-compressor by D. Rowan & Co Ltd, Glasgow. 1,950 I.H.P. — 11 kts.
18.12.1936 launched and 1.1937 completed by Wm Hamilton & Co Ltd, Port Glasgow. 1937: Temple S.S. Co Ltd (Lambert Bros Ltd), London, renamed TEMPLE YARD. 1951: Cia Nav Sevillana S.A., Panama (Capeside S.S. Co Ltd, London), renamed FLISVOS. 1952: Taiheyo Kaiun K.K. (Kyokuyo Hogei K.K.), Tokyo, renamed HOYO MARU. 1960: re-engined, oil engine by Kobe Hatsudoki Seizoshi, 3,800 I.H.P. — 10 kts. 1967: Dowa Kaiun K.K., Tokyo. 1969: Ta Fu Sg Co S.A., Panama (Ednasa Co. Ltd, Hong Kong), renamed YUAN TUNG. 1970: sold Chi Shun Hwa Steel & Iron Works Ltd, Kaohsiung, and breaking up commenced 1.5.1970.
Scrapped:
*GLITRA* (ex Greldale, Sterndale), 2,925/10. South Georgia Co Ltd (Chr Salvesen & Co.), Leith (£5,750). 1937 — West of Scotland Sbkg Co. Ltd, Troon. Arrived 27.3.1937.
*TRIAINA* (ex Meropi, Polgowan, Macedonia), 4,358/00. G. Ch. Lemos and M. A. Georgilios, Chios (£8,950). 1937 — Arnott, Young & Co Ltd, Dalmuir (£6,537).
*CHARALAMBOS* (ex Athena, Hermiston), 4,389/01. Ant Yannoulatos Sons, Piraeus (£9,190). 1937 — P. & W. MacLellan Ltd, Bo'ness (£6,584).

**NEREUS,** launched as DARLYON. *(Skyfotos)*

**Cumbrae Sg Co Ltd**
**(DARLYON)** Price/Loan: £88,000/£88,000
Details as DARLENY, except 3,122n and machinery by Rankin & Blackmore Ltd, Greenock.
24.8.1937 launched as DARLYON, sold and 10.1937 completed as NEREUS by Wm Hamilton & Co Ltd, Port Glasgow. 1937: Nereus S.N. Co Ltd, Athens (Hadjilias & Co Ltd, London). 9.10.1964: arrived Osaka and breaking up commenced 12.1964 by Nichimen & Co, Mihara.
Scrapped:

*VILLE DE DUNKERQUE* (ex Sebeto, Rovato, War Amazon), 3,190/18. Cie des Bateaux a Vapeur du Nord, Paris. 1937 - P. & W. MacLellan Ltd, Bo'ness.

*KUHRS* (ex Mette Jensen, Ausable, Laura), 3,160/01. M. Kalnins, Riga (£6,550). 1937 - P. & W. MacLellan Ltd, Bo'ness.

*BAKAR* (ex Chesham, Singapore), 4,295/06. Parabrodarsko Akcionarsko Drustvo 'Progress,' Susak (£8,500). 1938 — Smith & Houston, Port Glasgow.

Notes: The CHARALAMBOS had twice been sold to Italian breakers by her Greek owners, but in both cases the sale was not completed. She then passed to the Carrick Sg Co. The reason was probably delay in settlement of purchase price by the Italian breakers due to Italian exchange control.
The sale of the DARLENY and DARLYON at £135,000 each indicates the rapid rise in values between the dates they were ordered and completed.

**A. L. Duggan & Co, Bristol**
**Ald Sg Co Ltd**
**CASTLE COMBE** Price/Loan: £16,045/£16,045
O.N. 163872. 454g. 232n. 570d. 155.0 x 27.6 x 8.6 feet.
Oil engine (aft) by Ruston & Hornsby Ltd, Lincoln. 500 B.H.P. —9.5 kts.
23.9.1936 launched and 12.1936 completed by Charles Hill & Sons Ltd, Bristol. 1952: Plym Sg Co Ltd, Plymouth, renamed ALFRED PLYM. 1955: lengthened, now 623g. 349n. 982d. 190.0ft.

**CASTLE COMBE.** *(G. E. P. Brownell)*

1958: Capt. S. Lucchi, Venice, renamed COSTANZA. 1960: Mario Attanasio, Naples, renamed GIANNI ATTANASIO. 1970: Thalassia S.p.A., Genoa. 1979: broken up at Brindisi.
Scrapped:
*SUSA* (ex Echo), 925/83. Otto Behnke, Danzig (£1,950). 1936 — Danziger Werft u. Eisenbau Werkstatten A.G., Danzig (£1,650).
Proposed a second vessel, of 470g. 630d. from Hill at a cost of £18,350.
Application abandoned owing to inability to obtain tonnage by 25.2.1937.

**Robert J. Dunlop, Glasgow**
**Lomond Sg Co Ltd**
**DUNKELD** Price/Loan: £92,000/£92,000.
O.N. 164118. 4,944g, 2,994n, 9,200d. 418.2 x 55.2 x 25.5 feet.
Doxford oil engine by builders. 1,800 B.H.P. — 11 kts.

**DUNKELD.** *(A. Duncan)*

2.4.1937 launched and 5.1937 completed by Barclay, Curle & Co Ltd, Glasgow. 1943: Stanhope S.S. Co Ltd (J. A. Billmeir & Co Ltd), London. 1945: renamed STANKELD. 1951: Westralian Farmers Ltd, London, renamed SWANBROOK. 1956: China Sg Co Ltd (John Manners & Co Ltd), Hong Kong, renamed SYDNEY BREEZE. 1958: South Breeze Nav Co Ltd (John Manners & Co Ltd), Hong Kong. 1964: San Fernando S.S Co S.A., Panama (John Manners & Co Ltd, Hong Kong), renamed SAN ERNESTO. 1966: Oriental Trader Nav Co S.A., Panama (China Pacific Nav Co Ltd, Hong Kong), renamed CATHAY TRADER. 1968: Renown Sg Cpn S.A., Panama (Atlantic Sg & Tg Co Ltd, Hong Kong), renamed RENOWN TRADER. 8.1.1970 arrived Hong Kong, breaking up by Cheong Wah commenced 28.1.1970.
Scrapped:
*AIMILIOS* (ex Nicolaos Zafirakis, Cambrian King, Ullapool), 3,638/98. G. Grohmann, Athens (£7,500). 1936 — Frank Rijsdijk's Industrieele Ondernemingen, Hendrik-Ido-Ambacht (£5,600).
*MARIE SCHRODER* (ex Hammerburg, Stella Maris, Beta, Sirius), 742/89. Leth & Co, Hamburg. 1936 — Walter Ritscher, Hamburg.
*HEDWIG* (ex Hedwig Fischer, Mary, Praga, Hedwig Fischer, Pollux, Byzanz, Godrevy), 1,827/92. Leth & Co, Hamburg. 1936 — Walter Ritscher, Hamburg.
*TIBER*, 1,343/99. Det Forenede Dampskibsselskab, Copenhagen (£2,750). 1936 — Petersen & Albeck A/S, Copenhagen (£2,110).
*AGII VICTORES* (ex Siciliano, Capelbay, Cairnloch), 1,639/95. Const Caroussis, Chios (£3,150). 1937 — Thos W. Ward Ltd, Pembroke Dock. Arrived 4.9.1936 (£2,260).
Notes: The MARIE SCHRODER and HEDWIG were bought with the NOORDSEE (see Thomas Dunlop & Sons) for £7,500 and sold for scrapping for £6,000.
Robert Jack Dunlop was a brother of Sir Thomas Dunlop and a partner in Thomas Dunlop & Sons (which see).

**Thomas Dunlop & Sons, Glasgow**
**Queen Line Ltd**
**QUEEN MAUD** Price/Loan: £90,950/£90,950.
O.N. 164101. 4,976g. 3,038n. 9,200d. 422.8 x 54.2 x 26.1 feet.
Doxford oil engine by builders. 1,800 B.H.P. — 11 kts.
29.9.1936 launched and 11.1936 completed by Wm Doxford & Sons Ltd, Sunderland.
5.5.1941: torpedoed by U38 in 7.54N 16.41W. Cardiff for Freetown and Alexandria, coal and government stores. 39 crew, 1 lost.

**QUEEN ADELAIDE** Price/Loan: £88,500/£88,500.
O.N. 164093. 4,933g. 2,993n. 9,200d. 418.2 x 55.2 x 25.5 feet.
Doxford oil engine by builders. 1,800 B.H.P. — 11 kts.
30.7.1936 launched and 9.1936 completed by Barclay, Curle & Co. Ltd, Glasgow. 1951: Westralian Farmers Ltd, London, renamed SWANHILL. 1956: China Sg Co Ltd (John Manners

**QUEEN MAUD,** sole Doxford built member of the Dunlop fleet. *(E. N. Taylor)*

& Co Ltd), Hong Kong, renamed LONDON BREEZE. 1960: Cambay Prince S.S. Co Ltd (John Manners & Co Ltd), Hong Kong, renamed DAIREN. 1964: Haitong Steamships & Tg Co Ltd (Hornbeam Co Ltd), Hong Kong. 1966: Lanena Sg Co Ltd, Hong Kong (T. Engan, Manila), renamed AGATE. 1966: San Raimundo Cia Nav S.A., Panama (T. Engan, Manila). 1967: Express Trader Sg Co S.A., Panama (Yusang Sg Co, Hong Kong), renamed YU LEE. 21.3.1969: arrived Kaohsiung and breaking up by Lung Shi Steel & Iron Works Ltd commenced 20.6.1969.

**QUEEN VICTORIA** Price/Loan: £88,500/£88,500.
O.N. 164099. Details as QUEEN ADELAIDE except 4,937g.
15.9.1936 launched and 11.1936 completed by Barclay, Curle & Co Ltd, Glasgow. 28.6.1942: torpedoed by Japanese submarine I-10 in 21.15S 40.30E. Table Bay for Aden, government stores. 48 crew, no survivors.
Scrapped:

*BOMARSUND* (ex Cooee, Neumunster), 4,337/07. A/B Naxos Prince (R. Mattson), Helsinki. 1936 — Ghent (£6,700).

*COMTE DE FLANDRE* (ex Akropolis, Alpine Range, Kintail), 3,621/07.

*COMTESSE DE FLANDRE* (ex Bankdale), 3,851/07.

*ROI LEOPOLD* ((ex Karnak), 3,174/06.

The above three vessels sold by Cie Dens-Ocean S.A., Antwerp (£19,000). 1936 — Frank Rijsdijk's Industrieele Ondernemingen, Hendrik-Ido-Ambacht (£14,750).

*PEMBROKESHIRE,* 7,808/15. Glen Line Ltd, London (£19,750). 1936 — Danzig breakers (£14,700).

*STORK,* 2,029/04 (part). General Steam Nav Co Ltd, London (£4,000). 1936 — Hughes Bolckow Sbkg Co Ltd, Blyth. Arrived 29.3.1936 (£3,100).

*WYTHEVILLE,* 6,098/19. American Diamond Lines Inc, New York (£9,500). 1936 — Metal Industries Ltd, Rosyth (£8,750). Demolition began 27.5.1936.

**PEMBROKESHIRE** at Erith, 4 May 1935. *(World Ship Photo Library)*

**Cadogan S.S. Co Ltd**

**QUEEN ANNE** Price/Loan: £91,500/£91,500

O.N. 164105. Details as QUEEN ADELAIDE, except 4,937g.

4.11.1936 launched and 1.1937 completed by Barclay, Curle & Co Ltd, Glasgow. 10.2.1943: torpedoed by U509 in 34.53S 19.51E. Manchester & Table Bay for Aden, Alexandria & Beirut, general. 44 crew, 5 lost.

Scrapped:

*STORK* (part) (see above).

*HILLCROFT* (ex Sheba), 2,268/12. Westbourne Sg Co Ltd (G.N. McNeill Ltd),Cardiff. 1936 — Thos W. Ward Ltd, Briton Ferry. Arrived 5.2.1936 (£2,800).

*VIRGO* (ex Gwynmead, Elgin), 3,835/06. Red A/B Iris (E. Erikson), Mariehamn (£7,600). 1936 — Thos W. Ward Ltd, Grays (£6,000).

*NOORDSEE* (ex Klaipeda, Hochsee, Johannes Tiemann, Balmore), 1,293/90. W. Schuchmann, Bremerhaven. 1936 — Walter Ritscher, Hamburg.

*DANIA* (ex Dannebrog, Paul Pagh, Theodoros, North Flint), 2,247/88. Nordjyllands Kul-Kompagni A/S, Norresundby (£4,250). 1936 — Petersen & Albeck A/S, Copenhagen (£3,100).

Notes.- See also Robert J. Dunlop.

The NOORDSEE was bought with the MARIE SCHRODER and HEDWIG. See Robert J. Dunlop for details.

The HILLCROFT was initially proposed for scrapping by Westbourne Sg, but was then sold to the Rising Sun Nav Co in 1935 for £3,200. When no recommendation was made on the Rising Sun Co application for a loan she was again sold, to Cadogan S.S. Co in 1936 for £4,050.

**E.R. Management Co Ltd, Cardiff**

**Nailsea S.S. Co Ltd**

**NAILSEA MOOR** Price/Loan: £89,645/£88,625.

O.N. 162123. 4,926g. 2,948n. 8,950d. 420.3 × 56.0 × 25.4 feet.

White C 4-cyl plus LP turbine by White's Marine Engineering Co Ltd, Newcastle. 1,750 I.H.P. — 10 kts.

12.6.1937 launched and 9.1937 completed by Bartram & Sons Ltd, Sunderland. 1948: W.I. Radcliffe S.S. Co Ltd & Wynnstay S.S. Co Ltd (Evan Thomas Radcliffe & Co), Cardiff. 1949: renamed LLANWERN. 1951: Inui Kisen K.K., Kobe, renamed KENKON MARU. 1955: re-engined, T 3-cyl by Harima Zosensho, Aioi. 2,000 I.H.P. — 10.5 kts. 1961: Yamata Gyogyo K.K., Tokyo, renamed FUJISAN MARU. 1964: Nihon Kokai Gyogyo K.K., Tokyo, renamed RENSHIN MARU. 1969: broken up at Kaohsiung.

Scrapped:

*RODSKAR* (part) (see Notes).

*AIKLEAF* (ex Bedeburn, PLM16, Kifunezan Maru), 3,122/16. Dunelm Sg Co Ltd (Germain, Dixon & Co Ltd), Newcastle (£6,250). 1936 — Metal Industries Ltd, Rosyth (£4,500). Demolition began 19.2.1936

*RAMSES* (ex Maurice Colignon, Carama, Virginia, Euxinia), 3,773/99. A. Klat, Alexandria (£7,650). 1936 — Metal Industries Ltd, Rosyth (£6,100). Demolition began 8.7.1936

*KINGSBURY* (part) (see Notes).

NAILSEA MOOR. *(Tyne & Wear Archives)*

**NAILSEA MANOR** Price/Loan: £94,300/£94,300.

O.N. 162126. Details as NAILSEA MOOR, except 2,946n.

21.9.1937 launched and 12.1937 completed by Bartram & Sons Ltd, Sunderland. 10.10.1941: torpedoed by U126 in 18.45N 21.18W. Penarth for Freetown & Suez, stores. 41 crew, all saved.

Scrapped:

*NAILSEA VALE* (ex Clan Macbeth), 4,647/13. Clan Line Steamers Ltd (Cayzer, Irvine & Co Ltd), London (£16,250). 1938 — Arnott, Young & Co (Sbkg) Ltd, Dalmuir (£11,000).

*NAILSEA TOWER* (ex Chinkoa, Arabistan), 5,222/13. British India S.N. Co Ltd, London (£20,000). 1938 — Arnott, Young & Co (Sbkg) Ltd, Dalmuir (£12,000).

Notes:- See also Evans & Reid Investment Co.

The balance of the RODSKAR was credited to the Barry Sg, Co, which see under B. & S. Sg.

The KINGSBURY was also credited to Bantham Sg Co, which see under Evans & Reid Investment Co for details.

The HELLENIC (4,404/09) was also reported bought by the Nailsea S.S. Co in 1936 for scrap & build, but was renamed NAILSEA BELLE and sold in 1937 for further trading as the ST USK.

**E. Edwards & Son**

Proposed a tramp of 5,200g, 8,250d, costing £75,000.
No recommendation made.

**T.E. Evans & Co Ltd, London**

**ASHANTI** Price/Loan: £16,125/£16,125 (interest at 2.75 per cent, deferred 1 year).
O.N. 164622. 534g. 274n. 600d. 183.6 × 27.1 × 8.0 feet.
Oil engine (aft) by Humboldt-Deutzmotoren A.G., Koln. 408 B.H.P. — 9.75kts.
28.3.1936 launched and 5.1936 completed by Goole Shipbuilding & Repairing Co Ltd, Goole.
1938: re-engined with oil engine by Nydqvist & Holm A/B, Trollhattan. 400 B.H.P. — 9.25 kts.
10.6.1944: torpedoed by E-boat (S177 or S178), 35 miles N of Normandy assault area. I.o.W. convoy assembly point for Normandy, cased petrol. 17 crew, no survivors.

**BENGUELA.** *(L. Dunn)*

**BENGUELA** Price/Loan: £16,375/£16,375 (interest at 2.75 per cent, deferred 1 year).
O.N. 164641. Details as ASHANTI.
25.4.1936 launched and 6.1936 completed by Goole Shipbuilding & Repairing Co Ltd, Goole.
1943: Metcalf Motor Coasters Ltd, London. 1946: renamed ELLEN M. 1969: D.P. Kalkassinas, Thessalonika, renamed ATHANASSIOS KALKASSINAS. 1972: El. Selimos & Co, Piraeus, renamed PLAKAS. 1974: H. Piatoulakis, Piraeus, renamed ASTERIAS. 1975: Eleftherios Selimos, Piraeus, renamed PSILORITIS. 1979: broken up.

**CABENDA** Price/Loan: £16,125/£16,125 (interest at 2.75 per cent, deferred 1 year).
O.N. 164672. Details as ASHANTI, except engine by Mirrlees, Bickerton & Day Ltd, Stockport. 420 B.H.P. — 9.75 kts.
8.6.1936 launched and 7.1936 completed by Goole Shipbuilding & Repairing Co Ltd, Goole.
28.2.1941: sunk by mine in 51.34N 3.54W. Shoreham for Briton Ferry, scrap iron. 12 crew, 1 lost.

**LOANDA** Price/Loan: £16,375/£16,375 (interest at 2.75 per cent, deferred 1 year).
O.N. 164698. Details as CABENDA.
8.7.1936 launched and 8.1936 completed by Goole Shipbuilding & Repairing Co Ltd, Goole.
1953: Metcalf Motor Coasters Ltd, London, renamed MONICA M. 1955: re-engined with oil engine by Blackstone & Co Ltd, Stamford. 600 B.H.P. 1960: Wimaisia Sg Co Ltd, Glasgow (T.J. Metcalf, London). 1971: Stefanos Sikalos (E.V. Velkiadis), Piraeus, renamed PANORMITIS. 1975: M. Zerva & Ors, Piraeus, renamed AGIOS NICOLAOS. 9.2.1977: cargo shifted in heavy weather off Haifa and foundered. Limassol for Hodeidah, general.

**PURLEY OAKS.** *(A. Duncan)*

Scrapped:
*PURLEY OAKS* (ex Morawitz, Kermoor, Morawitz), 4,288/07. T.E. Evans & Co Ltd, London. 1936 — Metal Industries Ltd, Rosyth (£6,000). Demolition began 29.1.1936.
Proposed a 600g, 800d, vessel by Goole Sbg for £18,500.
Application abandoned due to inability to obtain tonnage by 25.2.1937.
Notes:- The original application was for five 500d vessels with a 7,000d vessel for scrapping which was employed in trade to the Mediterranean and Plate. No recommendation was made as the proposed vessels appeared to be excluded by the coasting restriction in the Act. Evans amended the application first to four of 500d and one of 400d, then to four of 600d which they satisfied the Committee would be suitable for Mediterranean trade. The abandoned application was probably the ANDONI (678/37).

**Evans & Reid Investment Co Ltd, Cardiff**
**Bantham S.S. Co Ltd**
**NAILSEA COURT** Price/Loan: £82,300/£77,340 (interest at 2.75%, not deferred).
O.N. 162113. 4,946g. 2,914n. 8,950d. 420.3 × 56.0 × 25.4 feet.
C 4-cyl plus LP turbine by White's Marine Engineering Co Ltd, Newcastle. 1,750 I.H.P. — 10 kts.
9.6.1936 launched and 8.1936 completed by Bartram & Sons Ltd, Sunderland. 10.3.1943: torpedoed by U229, convoy SC121, 58.45N 21.57W. Beira, Table Bay & New York for London, general. 46 crew & 2 passengers, 43 crew & 2 passengers lost.
Scrapped:
*CAPE ORTEGAL*, 4,896/11. Lyle Sg Co Ltd, Glasgow (£8,250). See note on scrapping.
*ANCHORIA* (part), 6112/11. T. & J. Brocklebank Ltd, Liverpool (£8,500). 1936 — Osaka £12,100).

**NAILSEA MEADOW** Price/Loan: £82,950/£74,655 (interest at 2.75 per cent, not deferred).
O.N. 162118. Details as NAILSEA COURT, except 4,962g, 2,943n, 1,760 I.H.P.
17.12.1936 launched and 2.1937 completed by Bartram & Sons Ltd, Sunderland. 1941: Nailsea S.S. Co Ltd, Cardiff. 19.3.1941: damaged by bomb in Victoria Dk, London. 2 killed. 11.5.1943: torpedoed by U196 in 32.4S 29.13E. Hampton Roads, New York, Trinidad, Bahia, Rio de Janeiro & Table Bay for Durban, Bombay & Karachi, war materials & general. 46 crew, 2 lost.

**NAILSEA COURT.** *(World Ship Photo Library)*

Scrapped:

*ANCHORIA* (part), see above.

*MARIETTA* (ex W.I. Radcliffe, Clarissa Radcliffe), 6,042/13. N. Eustathiou & Co, Piraeus (£7,700). 1937 — Osaka (£9,500).

*KINGSBURY* (ex Ilwen) (part), 4,072/04. Alexander Sg Co Ltd (Capper, Alexander & Co), London (£8,000). 1936 — West of Scotland Sbkg Co Ltd, Troon. Arrived 24.8.1936 (£5,500).

Notes:-See also E. & R. Management Co Ltd.

CAPE ORTEGAL passed through the hands of Heston Sg Co (which see) before being purchased by Bantham S.S. Co. Although sold to breakers (Metal Industries Ltd for £10,750 in 1939) she was taken over in 1939 and expended as a blockship at Scapa Flow.

The ANCHORIA was initially sold to Italian breakers for £8,750, but the deal fell through and she was sold to Japanese breakers.

**Frazer & Co, Newcastle**

Proposed a 3,100g, 4,750d, tramp by Laing costing £70,000. Application withdrawn.

Notes:- A firm of ship chandlers. James Frazer was also a director of Burnett S.S. Co and Clive Sg Co, both of which see.

**ARABIAN PRINCE** in war livery, 21 January 1944. *(World Ship Photo Library)*

**Furness, Withy & Co Ltd, London**

**Prince Line Ltd**

**ARABIAN PRINCE** Price/Loan: £65,000/£65,000.
O.N. 164728. 1,960g. 1,035n. 3,000d. 296.3 × 44.2 × 16.4 feet.
T 3-cyl by J. G. Kincaid & Co Ltd, Greenock. 1,700 I.H.P. — 12 kts.
2.9.1936 launched and 10.1936 completed by Wm Hamilton & Co Ltd, Port Glasgow. 4.4.1959: arrived Rotterdam and broken up by N.V. Holland Scheeps en Maschinenhandel, Hendrik-Ido-Ambacht.

**SYRIAN PRINCE** Price/Loan: £65,000/£65,000.
O.N. 165353. 1,990g. 1,003n. 3,000d. 296.5 × 44.2 × 16.4 feet.
T 3-cyl by Richardsons, Westgarth & Co Ltd, Hartlepool. 1,850 I.H.P. — 12 kts.
1.10.1936 launched and 12.1936 completed by Furness Shipbuilding Co Ltd, Haverton Hill-on-Tees. 1959: Cia Mar Med Ltda, San Jose (D. Th. Petropoulos, London), renamed SUNNY MED. 1964: Glyfada Seafaring Cpn, Panama (Methenitis & Co, Piraeus), renamed DINOS. 25.10.1969: laid up; breaking up commenced 23.4.1971 by Giuseppe Riccardi, Vado.

**SAILOR PRINCE.** *(World Ship Photo Library)*

Scrapped:

*SAILOR PRINCE* (ex Glendevon), 4,317/07. Rio Cape Line Ltd (Furness, Withy & Co Ltd), London. 1936 — Metal Industries Ltd, Rosyth (£6,500). Demolition began 24.6.1936.

*STUART PRINCE* (ex Glendhu), 4,129/05. Rio Cape Line Ltd (Furness, Withy & Co Ltd), London. 1936 — Danziger Werft u. Eisenbau Werkstatten A.G., Danzig (£6,100).

Proposed two further 1,900g, 3,000d, vessels, one each from Furness and Hamilton costing £62,000 each.

Applications withdrawn.

Notes:- The two further proposals were the CYPRIAN PRINCE (1,988/37) and PALESTINIAN PRINCE (1,960/36).

Also proposed for scrapping were the LONDON CITIZEN (ex Valemore) (5388/18) and LONDON EXCHANGE (ex Parisiana) (5,415/21). The LONDON CITIZEN was sold to the Lancashire Sg Co (which see) and credited in their scrap & build programme, whilst the LONDON EXCHANGE was sold to Ben Line for further trading as the BENRINNES.

**CROMARTY FIRTH.** Still afloat as the Greek MARIA PRECA (see page 15).
*(World Ship Photo Library)*

**G.T. Gillie & Blair Ltd, Newcastle**
**Northern Coasters Ltd**

**CROMARTY FIRTH** Price/Loan: £17,880/£17,880.
O.N. 161603. 538g. 275n. 725d. 161.5 × 28.1 × 9.9 feet.
Polar oil engine (aft) by British Auxiliaries Ltd, Glasgow. 420 B.H.P. — 10 kts.
12.1.1937 launched and 2.1937 completed by John Lewis & Sons Ltd, Aberdeen. 1954: Firth Sg Co Ltd (G.T. Gillie & Blair Ltd), Newcastle. 1957: James H. Henderson (P.D. Hendry & Sons), Glasgow, renamed HERRIESDALE. 1962: Aghios Georgios Sg Enterprises Ltd (C.G. Ventouris),

**SAPPHO.** *(World Ship Photo Library)*

Piraeus, renamed GEORGIOS VENTOURIS. 1979: Prekas Sg, Piraeus, renamed MARIA PREKA. Still in service.
Scrapped:
*SAPPHO,* 1,275/00. Bristol S.N. Co Ltd, Bristol (£2,500). 1936 — Thos W. Ward Ltd, Pembroke Dock. Arrived 19.5.1936 (£1,900).

**Gripay & Co Ltd, Cardiff**
Proposed a Burntisland single decker of 4,600g, 8,000d, costing £61,500.
No recommendation made.

**Hessler & Co, West Hartlepool**
**Hartlepool Seatonia S.S. Co Ltd**
**Swift S.S. Co Ltd**
Proposed two 5,580g, 9,000d, single deck, 3-island tramps from Gray, costing £81,000 each.
No recommendation made.

**Charles Hill & Sons, Bristol**
**Bristol City Line of Steamships Ltd**
Enquired about a loan for a 2,950g, 5,000d, liner by Hill, price £70,000.
Application withdrawn.

**W.S. Hinde, London**
**Penhill Sg Co Ltd**
Applied for a loan for a 4,290g, 7,555d, single deck self-trimmer by Gray, costing £61,000.
No recommendation made.
Notes:-The loan was not recommended due to lack of capital, which was consequently increased from £10,000 to £15,000. Despite this the application was not proceeded with.

**R.I. James, Newcastle**
**White Sg Co Ltd**
**BIDDLESTONE** Price/Loan: £75,700/£75,700.
O.N. 161606. 4,910g. 2,953n. 8,000d. 401.9 × 53.9 × 26.6 feet.
C 4-cyl plus LP turbine by White's Marine Engineering Co Ltd, Newcastle. 1,660 I.H.P — 10 kts.
10.5.1937 launched and 7.1937 completed by Short Bros Ltd, Sunderland. 1940: Clarissa Radcliffe S.S. Co Ltd (Evan Thomas Radcliffe & Co), Cardiff, renamed LLANCARVAN. 30.5.1943: bombed by aircraft 2 miles S of Cape St Vincent. Lisbon for Gibraltar, coal. 49 crew, all saved.

**BIDDLESTONE.** *(National Maritime Museum)*

Scrapped:
*DUNAFRIC* (ex Lidvard, Songvard, Lidvard, Pangan), 3,489/09. Bank Line Ltd (A. Weir & Co), London (£6,750). 1936 — Thos W. Ward Ltd, Inverkeithing. Arrived 17.7.1936 (£4,840).
*GLENAMOY,* 7,306/16 (part). Glen Line Ltd, London (£15,400). 1936 — Metal Industries Ltd, Rosyth (£11,500). Demolition began 18.11.1936
Notes:- This company was formed in 1933 to demonstrate and promote the White compound engine, their first vessel being the ADDERSTONE (ex Boswell) (5,255/20) which, with the BIDDLESTONE under construction, was sold to Norwegian owners in 1937.
The company had a capital of £15,000 shared between W.A. White (£11,000) and R.I. James (£4,000).

**Richard W. Jones & Co, Newport**
**Uskside S.S. Co Ltd**
**USKSIDE** Price/Loan: £47,953/£45,000.
O.N. 162143. 2,706g. 1,563n. 4,500d. 314.0 × 45.2 × 21.7 feet.
T 3-cyl by D. Rowan & Co Ltd, Glasgow. 830 I.H.P. — 9.5 kts.
26.11.1936 launched and 1.1937 completed by Burntisland Shipbuilding Co Ltd, Burntisland. 1.8.1943: bombed by aircraft, burnt out and sunk alongside quay at Palermo, army stores

**USKSIDE** as the Italian TESEO (see also page 76). *(L. Dunn)*

including petrol in drums and some ammunition. 44 crew, 1 killed. T.L. 1946: wreck sold, lifted and repaired. 1946: Soc Ligure di Armamento, Genoa, renamed TESEO. 24.12.1955: aground at Yamaniguey, near Baracoa. Yamaniguey for Baltimore, chrome ore. 29.12.1955 broke in two.
Scrapped:
*ENA G* (ex Yoseric, War Parrot), 5,240/18. Bank Line Ltd (A. Weir & Co), London (£10,500). 1936 — John Cashmore Ltd, Newport (£7,250).
Notes:- The ENA G was purchased from Bank Line in 1936 for £14,250 by the Glenfield Syndicate Ltd, London, and sold by them to Uskside S.S. Co for £10,500.

**T.L. Lees, West Hartlepool**
Enquired for a single deck 'easy trimmer' of 2,700g, 4,200d, costing £42-45,000.
No recommendation made.
Notes:- T.L. Lees, aged 37, was in charge of the chartering department of Smith, Hogg & Co Ltd, West Hartlepool, by whom he had been employed since 1919.

**C. Lochen & Co Ltd, Newcastle**
**Bjønnes & Co**
Proposed a 2,029g, 3,500d, timber carrier by Swan Hunter costing £43,250.
No recommendation.
Offered the Norwegian BRASK (4,079/11) for scrapping.
Notes:- This was an application on behalf of a Norwegian company forming a British subsidiary.

**Lykiardopulo & Co Ltd, London**
Proposed a 5,000g, 9,200d, vessel from Doxford costing £92,000.
No recommendation.

**McCowen & Gross Ltd, London**
**DERRYMORE** Price/Loan: £117,200/£117,200.
O.N. 166384. 4,799g. 2,822n. 9,500d. 414.0 × 56.9 × 25.7 feet.
Oil engine by Wm Doxford & Sons Ltd, Sunderland. 2,100 B.H.P. — 11.5 kts.
18.1.1938 launched and 4.1938 completed by Burntisland Shipbuilding Co Ltd, Burntisland. 13.2.1942: torpedoed by Japanese submarine I-25 in 5.18S 106.20E, Singapore for Batavia, aircraft & military stores. 36 crew & 209 R.A.A.F. personnel, 9 R.A.A.F. personnel missing.
Scrapped:
*HEATHCOT* (ex Clan Macbride), 4,922/12. Clan Line Steamers Ltd (Cayzer, Irvine & Co Ltd), London. 1938 — Osaka (£18,000).
*MOORCOT* (ex Clan Mackellar), 6,387/13 (part). Clan Line Steamers Ltd (Cayzer, Irvine & Co Ltd), London. 1938 — Osaka (£19,250).
A second application was made for a similar vessel by Burntisland costing £117,200.
No recommendation for loan made.
Notes:- McCowen & Gross Ltd was formed in 1936, the SEAPOOL, renamed PILCOT, being their first vessel. When applying for the above loan their capital of £10,000 was deemed insufficient, so it was increased to £15,000 to satisfy the Ships Replacement Committee.
The CLAN MACBRIDE and CLAN MACKELLAR were purchased in 1936 for £37,000.

The balance of the MOORCOT tonnage under the Scrap & Build Scheme was utilised by the Springwell Sg Co (which see).
The second application was probably the DERRYNANE (4,896/38). The PILCOT (4,549/13) was offered for scrapping, but was sold to Estonia in 1939 for further trading as the VAPPER.

DERRYMORE. *(Henry Robb Ltd.)*

**J.D. McLaren & Co, London**
Proposed two 4,500g, 8,460d, tramps by Barclay, Curle, costing £72,350 each. No recommendation made.
Offered for scrapping the CHARLBURY (ex Sakharah) (6,030/06) and the MARIA (3,090/01), which were both credited for the St Quentin Sg Co (which see) scrap & build programme.

**G.H. McNeil Ltd, Cardiff**
**Westbourne Sg Co Ltd**
Proposed to build a vessel of 2,300g, 4,000d, by Burntisland, cost £40,000.
No recommendation made.
Offered their HILLCROFT for scrapping, but when no recommendation made for a loan she was sold to the Rising Sun Nav Co, who in their turn sold her to the Cadogan S.S. Co (both of which see).

**Wm Milburn & Co Ltd, Newcastle**
Proposed a steam tramp of 4,500g, 7,880d, costing £73,000.
Application abandoned owing to inability to obtain tonnage by 25.2.1937.

**Morel Ltd, Cardiff**
Proposed a 4,950g, 9,075d, vessel by Doxford, costing £100,000.
Application abandoned owing to inability to obtain tonnage by 25.2.1937.
Notes:- This vessel was probably the FOREST (4,998/37).

**Muir Young Ltd, London**
**Labrador Development Co Ltd, St Johns, Nfld**
Proposed a 5,100g, 8,000d, single-deck tramp by Burntisland, costing £81,000.
No recommendation made.
Notes:- J.O. Williams & Co Ltd, Cardiff, were behind the Labrador Development Co, which had a 99 year lease of 3,560 sq. miles of timberlands in Labrador.
Muir Young Ltd were reported as buyers in 1936 of the RINOS (ex Raisdale, Drammensfjord, Chiltern Range) (4,346/11), for scrap & build purposes. She was sold three months later to Japanese breakers but continued trading as the Chinese flag YONG SHYANG.

**J.C. Radcliffe, Cardiff**
**Heston Sg Co Ltd**
Proposed two 4,500g, 8,100d, tramps costing £70,000 each.
No recommendation made.
Offered for scrapping the WEARBRIDGE (4,014/11) and MEDIA (5,437/11).
Notes:- The Heston Sg Co was formed in 1935 with Halford Constant and John C. Radcliffe as directors.
The WEARBRIDGE was sold on to Anning Bros and scrapped against STARCROSS.
The MEDIA was sold to A. Lauro, Naples, for further trading as the VELOCE.
The CAPE ORTEGAL also passed through the hands of the Heston Sg Co before going to the Bantham S.S. Co (which see) for their scrap & build programme.

**Sir R. Ropner & Co Ltd, West Hartlepool**
**Ropner Sg Co Ltd**
Proposed a 4,956g, 9,170d, vessel from Doxford (619), at a cost of £91,500.
Application not proceeded with.
The COALBY (ex Kamouraska) (4,903/11), HOLTBY (3,681/09) and SPILSBY (3,673/10) were offered for scrapping.

Notes:- This vessel (619) was the MOORBY (4,992/36).
The COALBY was sold to the St Quentin Sg Co, and the SPILSBY to the Lancashire Sg Co (both of which see) and broken up in their scrap & build programmes.
The HOLTBY was sold to Italian breakers in 1935, but the sale was cancelled and a new sale arranged to Greek owners for further trading as the POSEIDONIA.

**Rose Line Ltd, Sunderland**
Proposed a 1,000g, 1,450d, vessel costing £45,000. No recommendation made.

**H.H. Ross, London**
Initially proposed two ships for the carriage of grain and cattle from Canada to the UK. Later proposed a 5,000g, 9,000d, tramp costing £85,000. No recommendation made.

**Runciman (London) Ltd**
**M. Embiricos & Co Ltd**
Proposed a 4,970g, 9,185d, tramp costing £95,000. No recommendation made.
Notes:- This vessel may have been the ELLIN (4,917/38).

**Sir William Reardon Smith & Sons Ltd, Cardiff**
**Reardon Smith Line Ltd**
**BRADFORD CITY** Price/Loan: £111,000/£88,554 (interest at 2.75 per cent).
O.N. 161622. 4,952g. 2,999n. 9,555d. 426.5 × 56.2 × 25.3 feet.
Oil engine by Wm Doxford & Sons Ltd, Sunderland. 2,900 B.H.P. — 12 kts.
5.5.1936 launched and 6.1936 completed by Furness Shipbuilding Co Ltd, Haverton Hill-on-Tees. 1.11.1941: torpedoed by U68 in 22.59N 9.49E. Mauritius & Table Bay for Freetown & U.K., sugar. 44 crew, all saved.
Scrapped:
*INDIAN CITY,* 6,221/20. Reardon Smith Line Ltd (Sir Wm Reardon Smith & Sons Ltd), Cardiff. 1935 — John Cashmore Ltd, Newport (£7,500).
*ORIENT CITY* (ex Cloughton), 5,622/11. Reardon Smith Line Ltd (Sir Wm Reardon Smith & Sons Ltd), Cardiff. 1935 — Cantiere Navale Scoglio Ulivi, Pola.

**CORNISH CITY.** *(E. N. Taylor)*

**Leeds Sg Co Ltd**
**CORNISH CITY** Price/Loan: £111,000/£105,000 (interest at 2.75 per cent).
O.N. 161624. Details as BRADFORD CITY except 2,984n.
16.9.1936 launched and 11.1936 completed by Furness Shipbuilding Co Ltd, Haverton Hill-on-Tees. 29.7.1943: torpedoed by U177 in 27.20S 52.10E. Lourenco Marques & Durban for Aden & Suez, coal. 43 crew, 37 lost.
Scrapped:
*FRANCISCO,* 6,272/10. Ellerman's Wilson Line Ltd, Hull (£9,250). 1935 — Cantiere Navale Scoglio Ulivi, Pola.
*SALIENT,* 3,879/05. Westoll Steamships Ltd (James Westoll Ltd), Sunderland (£6,000). 1935 — Hughes Bolckow Sbkg Co Ltd, Blyth. Arrived 10.7.1935 (£4,700).
Notes:- The ORIENT CITY and FRANCISCO were sold to Trieste breakers for £20,750.
In error the Ships Replacement Committee records list the ORIENT CITY as broken up at Newport and the INDIAN CITY at Trieste.

SALIENT. *(M. Lindenborn)*

**W.A. Souter & Co Ltd, Newcastle**
**Hebburn S.S. Co Ltd**

**HYLTON** Price/Loan: £95,937/£95,937.
O.N. 161601. 5,197g. 3,040n. 9,300d. 427.0 × 56.1 × 25.9 feet.
Oil engine by North Eastern Marine Engineering Co Ltd, Newcastle. 2,110 B.H.P. — 11 kts.
31.10.1936 launched and 1.1937 completed by Wm Pickersgill & Sons Ltd, Sunderland. 29.3.1941: torpedoed by U48, convoy HX115, in 60.2N 18.10W (600 miles W of Cape Wrath). Vancouver for Tyne, lumber & wheat. 36 crew, all saved.

Scrapped:

*DESPINA A PAPPA* (ex Elissavet V, Katina, Theone, Tweeddale), 4,649/04. A.G. Pappas, Chios (£8,500). 1936 — Vereinigte Stahlwerke, Düsseldorf (£7,600).

*CAPITAINE COULLON* (ex War Wasp), 1,325/17. Cie Generale Transatlantique, Paris (£2,025). 1936 — Haulbowline Industries Ltd, Queenstown (£1,600).

*VAGLIANO* (ex Hazel Branch, Bellgrano), 4,734/06. A. Lusi, London (£9,500). 1936 — Hughes Bolckow Sbkg Co Ltd, Blyth. Arrived 3.1.1936 (£8,000).

HYLTON. *(World Ship Photo Library)*

**Springwell Sg Co Ltd, London**

**SPRINGWEAR** Price/Loan: £27,580/£20,685.
O.N. 164652. 1,173g. 655n. 1,600d. 221.3 × 36.2 × 14.1 feet.
T 3-cyl (aft) by North Eastern Marine Engineering Co Ltd, Sunderland. 676 I.H.P. — 10 kts.
19.5.1936 launched and 6.1936 completed by Short Bros Ltd, Sunderland. 1937: High Hook Sg Co Ltd (Culliford & Clark Ltd), London, renamed HIGHWEAR. 1952: R. Muller (E. Ottens), Cuxhaven, renamed HANNES. 1953: Klondyke Sg Co Ltd, Hull, renamed BOSTONDYKE. 1962:

Ventouris Bros, Piraeus, renamed PANAGIA ODIGITRIA. 26.11.1964: Aground, engine room flooded outside Leningrad. Leningrad for Ghent, timber. Refloated, towed in 30.11.1964. C.T.L.
Scrapped:
*CORDELIA,* 2,506/89. Rederi A/B Cordelia (G.E. Sandstrom), Gothenburg (£5,350). 1937 — A/B Lindholmens-Motala, Gothenburg.

**SPRINGWOOD.** *(Tyne & Wear Archives)*

**SPRINGWOOD** Price/Loan: £27,500/£24,750.
O.N. 164679. Details as SPRINGWEAR, except 1,177g. 657n.
18.6.1936 launched and 7.1936 completed by Short Bros Ltd, Sunderland. 1937: High Hook Sg Co Ltd (Culliford & Clark Ltd) London, renamed HIGHWOOD. 1.7.1941: damaged by aircraft bomb at Barry, 1 killed. 1954: Granta S.S. Co Ltd (Witherington & Everett), Newcastle, renamed SPANKER. 7.8.1954: aground on breakwater during storm at entrance to Nieuwe Waterweg, Hook of Holland. Rotterdam for London, ballast.
Scrapped:
*GUNVALL* (ex Havtor, Bankchef Fasting, General Roberts), 1,494/82. Rederi A/B Vallde (J.A. Edvall), Oskarshamn (£3,000). 1936 — Thos W. Ward Ltd, Inverkeithing. Arrived 14.1.1936 (£1,875).
*MACEDONIA,* 1,561/81 (part). Rederi A/B Macedonia (I. Scheja), Stockholm (£3,350). 1936 — Liepajas Kari Ostes, Libau (£2,750).

**SPRINGWAVE** Price/Loan: £27,775/£24,997.10.0d.
O.N. 165335. Details as SPRINGWEAR, except 1,178g, 655n.
1.10.1936 launched and 11.1936 completed by Short Bros Ltd, Sunderland. 1937: High Hook Sg Co Ltd (Culliford & Clark Ltd) London, renamed HIGHWAVE. 30.1.1940: bombed by aircraft 1 mile NNE of Kentish Knock, 51.39.9N 1.41E. Hull for Lorient, coal. 18 crew, all saved. Wreck dispersed.
Scrapped:
*MACEDONIA* (part) — see above.
*JUNO* (ex Bollweiler, Juno, Lurline), 1,041/71. Rederi A/B Cordelia (G.E. Sandstrom), Gothenburg. 1937 — C. Dorkin & Co, Sunderland.
*GLENAMOY,* 7,306/16 (part). Glen Line Ltd, London. 1936 — Metal Industries Ltd, Rosyth (£11,500). Demolition began 18.11.1936.

**GLENAMOY.** *(World Ship Photo Library)*

**SPRINGTIDE** Price/Loan: £43,500/£43,500.
O.N. 165530. 1,579g. 893n. 2,400d. 251.4 × 40.3 × 17.9 feet.
T 3-cyl (aft) by North Eastern Marine Engineering Co Ltd, Newcastle. 750 I.H.P. — 10 kts.
22.6.1937 launched and 7.1936 completed by Short Bros Ltd, Sunderland. 1940: bought by R.N., mine destructor vessel. 1947: sold back to Springwell Sg Co Ltd. 1953: Eastboard Sg Ltd (Federal Commerce & Nav Co Ltd), Montreal, renamed EASTIDE. 1956: Aug Bolten, Hamburg, renamed SULLBERG. 1960: Cia Maritima Angelikana S.A., Panama (D.J. Papadimitriou Sons, London), renamed GRANNY MARIGO. 1961: Soc Alluminio Veneto per Azioni (Soc Abbruzzese di Nav per Az [SANA]), Venice, renamed GIUSEPPE RICCARDI. 1972: "Carbonavi" Soc di Nav S.p.A., Palermo. 1972: broken up at La Spezia.
Scrapped:
*RISALDAR,* 4,919/12. Asiatic S.N. Co Ltd, London (£17,500). 1937 — Metal Industries Ltd, Rosyth (£11, 500). Demolition began 4.8.1937.
*MOORCOT* (part) — see McCowen & Gross.

**SPRINGDALE** Price/Loan: £43,500/£43,500.
O.N. 165564. Details as SPRINGTIDE.
5.8.1937 launched and 9.1937 completed by Short Bros Ltd, Sunderland. 1940: bought by R.N., mine destructor vessel. 1947: sold back to Springwell Sg Co Ltd. 1953: Eastboard Sg Ltd (Federal Commerce & Nav Co Ltd), Montreal, renamed EASTDALE. 1955: Springwell Sg Co Ltd, renamed SPRINGDALE. 1959: Lynn Sg Co (Thos E. Kettlewell & Son Ltd), Hull. 18.6.1959: cargo shifted, capsized, sank in Gulf of Bothnia. 63.2N 19.39E. Munksund for London, timber.
Scrapped:
see SPRINGTIDE.
Notes:- The original application was for five 1,200g, 1,600d, motor tramps costing £24,000 each. No recommendation was made, the application was withdrawn and later re-submitted as above.

**SPRINGTIDE.** *(B. Feilden)*

**H.M.S. SPRINGTIDE** at Trincomalee, 1946. *(A. D. Townsend)*

**Stephens, Sutton Ltd, Newcastle**
**RIDLEY** Price/Loan: £91,150/£77,477.10.0d.
O.N. 161602. 4,993g. 3,061n. 9,185d. 423.5 × 54.2 × 26.1 feet.
Doxford oil engine by builders. 1,800 B.H.P. — 11 kts.
16.12.1936 launched and 1.1937 completed by Wm Doxford & Sons Ltd, Sunderland.
9.11.1940: on fire, abaondoned, 20.8N 29.36W. Glasgow for Pepel, ballast.
Scrapped:

*BULGARIAN* (ex Otto Kalthoff, Marie Menzell), 2,064/04. Westcott & Laurance Line Ltd (Ellerman Lines Ltd), London (£4,500). 1936 — Thomas Young, Sunderland (£3,200).

*UNION* (ex Rowanburn, Fernhill), 2,398/08. A/B Union S.S. Co (A/B Nielsen & Thorden O/Y), Helsinki. 1936 — South Stockton Sbkg Co Ltd, Stockton (£3,500).

*WHITEGATE* (ex Arsterturm), 5,067/11. Turnbull, Scott Sg Co Ltd (Turnbull, Scott & Co), London. 1937 — Van Heyghen Freres, Ghent (£9,850).

*SANTA AURORA* (ex Ibicuy), 5,529/19 (part). Eagle Oil & Sg Co Ltd, London (£8,900). 1937 — Eckhardt & Co A.G., Hamburg (£10,500).

SANTA AURORA. *(World Ship Photo Library)*

**Red 'R' S.S. Co Ltd**
**RUGELEY** Price/Loan: £87,900/£87,900 (interest at 2.75 per cent).
O.N. 161590. Details as RIDLEY, except 4,985g. 3,061n.
9.1.1936 launched and 2.1936 completed by Wm Doxford & Sons Ltd, Sunderland. 1948: Stylehurst Sg Co Ltd (Hadjilias & Co Ltd), London, renamed STYLEHURST. 1951: Rederi A/B Atos (Lundgren & Borjessons Rederier), Helsingborg, renamed ASPEN. 1960: Lombard Nav Co Inc, Panama (Gibson Sg Co Inc, Macao). 1966: Van Engineers Ltd, Hong Kong. 1967: Van Sg Co Ltd, Hong Kong, renamed METROPOLITAN. 4.6.1970: arrived Hong Kong, breaking up by Lee Sing commenced 17.6.1970
Scrapped:

*SUDBURY* (ex Southwaite), 3,632/04. Alexander Sg Co Ltd (Capper, Alexander & Co), London (£5,600). 1935 — South Stockton Sbkg Co Ltd, Stockton (£4,350).

*ELSISTON* (ex Muncaster Castle), 4,759/06. Elsiston Sg Co Ltd (Wm S. Miller & Co), Glasgow (£8,200). 1936 — Metal Industries Ltd, Rosyth (£6,650). Demolition began 25.12.1935.

*ROSTA* (ex Vossa, Clevedon, Chevalier St Georges, Vesta, Redemiral, Bansei Maru No 5), 1,332/19. Rederi A/B Rosta (E. Mortensen & O. Thue), Oslo (£2,000). 1936 — Grimstad Ophugnings Co A/S, Grimstad (£1,850).

*CARNARVONSHIRE,* 9,385/14 (part). Glen Line Ltd, London. 1936 — Amakusa Sangyo Kisen K.K., Osaka (£20,000).

**Whalton Sg Co Ltd**
**RILEY** Price/Loan: £87,900/£87,900 (interest at 2.75 per cent)
O.N. 161597. Details as RIDLEY.
4.6.1936 launched and 7.1936 completed by Wm Doxford & Sons Ltd, Sunderland. 1956: San Nicolaos Cpn, Monrovia (Angelos, Leitch & Co, London), renamed ANOULA A. 1967: Kien Ping S.S. Co. Ltd, Hong Kong (Tai Lai S.S. Co Ltd, Kaohsiung), renamed KIEN PING. 1968: broken up by Amakusa Sangyo Kisen K.K., Tokyo, completed 5.1968.
Scrapped:

*CARNARVONSHIRE* (part) — see above.

*CITY OF KHARTOUM* (ex Karroo), 6,132/13. Ellerman & Bucknall S.S. Co Ltd (Ellerman Lines Ltd), London. 1936 — Ditta Luigi Pittaluga, Genoa (£11,500).

**CITY OF KHARTOUM.** *(G. E. P. Brownell)*

*SANDGATE* (ex Alster), 3,687/06 (part). Redgate S.S. Co Ltd (Turnbull, Scott & Co), London. 1936 — Metal Industries Ltd, Rosyth (£5,500). Demolition began 1.4.1936.

*ALICE M. CRAIG,* 916/00. Hugh Craig & Co, Belfast (£2,500). 1936 — Edgar G. Rees, Llanelly (£1,625).

**ROTHLEY** Price/Loan: £89,400/£87,900 (interest at 2.75 per cent).
O.N. 161598. Details as RIDLEY, except 4,996g. 3,062n.
29.10.1936 launched and 12.1936 completed by Wm Doxford & Sons Ltd, Sunderland. 19.10.1942: torpedoed by U332 in 13.34N 54.34W. Durban for Trinidad & New York, ballast. 42 crew, 2 lost.
Scrapped:
see under RILEY.
Proposed another 4,960g, 9,200d, motor vessel by Doxford, costing £89,400. No recommendation made.

**ROTHLEY.** *(J. K. Byass)*

**Thomasson Sg Co Ltd, London**

**RIPLEY** Price/Loan: £90,900/£90,900.
O.N. 161599. Details as RIDLEY, except 4,997g. 3,063n.
3.10.1936 launched and 12.1936 completed by Wm Doxford & Sons Ltd, Sunderland. 1941: South American Saint Line Ltd (B & S. Sg Co Ltd), Cardiff. 12.12.1942: torpedoed by U161 in 0.35S 32.17W. Forcados, Duala & Takoradi for Trinidad & U.K., palm oil, mahogany & rubber. 41 crew, all saved.
Scrapped:

*SANDGATE* (part) — see above.

*HERAKLES* (ex Clan Mackinnon), 4,905/03. A/B Oceanfart (L. Krogius), Helsinki (£11,000). 1936 — P. & W. MacLellan Ltd, Bo'ness. Arrived 27.4.1936 (£8,450).

*SANTA AURORA* (part) — see above).

Notes:- CARNARVONSHIRE and ELSISTON were originally sold to Italian breakers (prices £17,500 and £8,400) but the contracts were cancelled. The third vessel for Whalton Sg Co was probably the RODSLEY, sold prior to completion to A/S J. Ludwig Mowinckels Rederi, Bergen, for £163,000 and completed as the TROMA (5,029/37). The ALICE M. CRAIG was sold to John Cashmore Ltd, Newport, for scrapping and resold to Edgar Rees, Llanelly.

**Stone & Rolfe Ltd, Llanelly**
**S. & R. Steamships Ltd**
Applied for a loan for a 950g, 1,340d, vessel from Burntisland costing £27,350.
The application was abandoned due to their inability to obtain tonnage for scrapping by 25.2.1937.
Note:- This vessel was the LOTTIE R (972/37).

HOPESTAR. *(National Maritime Museum)*

**Arthur Stott & Co Ltd, Newcastle**
**Wallsend Sg Co Ltd**
**HOPESTAR** Price/Loan: £87,500/£87,500 (interest at 2.75 per cent).
O.N. 161592. 5,267g. 3,192n. 9,800d. 416.8 × 57.4 × 27.0 feet.
Steam turbine by Parsons Marine Steam Turbine Co Ltd, Wallsend-on-Tyne. 2,000 S.H.P. — 11 kts.
22.1.1936 launched and 2.1936 completed by Swan, Hunter & Wigham Richardson Ltd, Wallsend-on-Tyne. 31.12.1948: posted missing at Lloyd's. Sailed Tyne 2.11.1948 for Philadelphia, ballast. In wireless communication 14.11.1948, in 42N 57W.
Scrapped:
*OTTO SVERDRUP,* 3,612/04. A/S D/S Otto Sverdrup (Bergh & Helland), Bergen. 1936 — Clayton & Davie, Dunston-on-Tyne.
*COCOSLEAF* (ex Bedefell, Fukuyo Maru), 3,718/19. Dunelm Sg Co Ltd (Germain, Dixon & Co Ltd), Newcastle. 1936 — Clayton & Davie, Dunston-on-Tyne.
*DIMITRIS N RALLIAS* (ex General Degoutte, Australstream, Daltonhall), 3,534/99. N.D. Rallias, Andros (£5,500). 1936 — Clayton & Davie, Dunston-on-Tyne.

**Novocastria Sg Co Ltd**
**HOPECASTLE** Price/Loan: £118,000/£118,000.
O.N. 161604. 5,178g. 3,130n. 9,700d. 418.9 × 57.4 × 25.6 feet.
Doxford oil engine by builders. 2,800 B.H.P. — 12.25 kts.
29.12.1936 launched and 5.1937 completed by Swan, Hunter & Wigham Richardson Ltd, Wallsend-on-Tyne. 28.10.1942: torpedoed by U509, convoy SL125, in 31.39N 19.35W. Cochin & Freetown for Mersey, magnesite, ilmenite, tea & general. 46 crew, 5 lost.
Scrapped:
*AUSTRALIA,* 7,551/12 (part). British India S.N. Co Ltd, London (£15,750). 1937 — Mitsuwa Shoji K.K., Osaka (£13,500).
*ATTIKOS* ( ex Athenic), 4,078/06. A. Lusi, London (£8,000). 1936 — South Stockton Sbkg Co Ltd, Stockton. (£6,000).

**Clive Sg Co Ltd**
**HOPECROWN** Price/Loan: £118,000/£118,000.
O.N. 161607. Details as HOPECASTLE, except 5,180g. 3,131n.
11.5.1937 launched and 7.1937 completed by Swan, Hunter & Wigham Richardson Ltd, Wallsend-on-Tyne. 1950: Clive Sg Co Ltd (Stephens, Sutton Ltd), Newcastle, renamed RADLEY. 1956: Rederi A/B Atos (Lundgren & Borjessons Rederier), Helsingborg, renamed ALSTERN. 1963: Santa Katerina Cia Nav S.A., Panama (G. Lemos Bros Co Ltd, London), renamed APOSTOLOS ANDREAS. 1967: (Transatlantic Seaways Ltd, London). 19.7.1967: aground on Formigas Bank, 18.52N 75.46W. Guantanamo for China, bagged raw sugar. C.T.L.
Scrapped:
*NUDDEA,* 7,928/19. British India S.N. Co Ltd, London (£15,500). 1937 — Osaka (£12,750).
*RIP* (ex St Agnes), 1,188/03. M. Goossens, Liege (£2,200). 1936 — John Cashmore Ltd, Newport (£1,750).
*AUSTRALIA* (part) — see above.

AUSTRALIA. *(B. Feilden)*

**Hopemount Sg Co Ltd**

Proposed a 5,100g, 9,600d, vessel from Swan Hunter or Barclay, Curle at a cost of £107,000. Application withdrawn.

Notes:- Arthur Stott & Co Ltd later became Stott, Mann & Fleming Ltd. The HOPESTAR was the first installation of a Parsons 'Simplex' turbine.

The Clive Sg Co was formed in 1936 with a capital of £15,000.

The principal shareholders in the Hopemount Sg Co were Swan Hunter & Wigham Richardson.

The Wallsend Sg Co was owned by Swan Hunter & Wigham Richardson, the Parsons Marine Steam Turbine Co Ltd and the Wallsend Slipway & Engineering Co Ltd.

The vessel proposed by the Hopemount Sg Co may have been the HOPEPEAK (5,179/38).

**Walter Vaughan (Cardiff) Ltd, Cardiff**
**Rising Sun Nav Co Ltd**
**Good Hope Sg Co Ltd**

Proposed a vessel of 2,060g, 3,500d, from Burntisland costing £42,000.

No recommendation made.

The HILLCROFT was purchased from Westbourne Sg Co in 1935 for scrapping, and resold to Cadogan S.S. Co (both of which see).

**Walford Lines Ltd, London**

Proposed a 315g, 320d, tramp by Goole or Burntisland, costing £13,000.

No recommendation made.

**David Williamson Ltd, London**

Proposed three shallow draft motorships from either Hawthorn Leslie or Henry Robb, of 350g, 520d, £17,500: 260g, 350d, £11,500: 225g, 300d, £10,000. No recommendations made.

**Wing Management Co Ltd, Cardiff**
**Wing Line Ltd**
**Wavelet Sg Co Ltd**

Proposed a 5,620g, 9,000d, tramp by Laing costing £97,500. No recommendation made.

HOPECROWN. *(L. Dunn)*

# SHIPPING LOAN — FLEET LISTS

**Allan, Black & Co, Sunderland**
**Albyn Line Ltd**
Grant paid on THISTLEGORM (4,900/40).

**Ambrose, Davies & Matthews Ltd, Swansea**
**Brynymore S.S Co Ltd**
**DAN-Y-BRYN** Price/Loan: £123,000/£110,000
O.N. 167382. 5,117g. 3,034n. 9,940d. 420.0 x 58.0 x 26.3 feet.
T 3-cyl by D. Rowan & Co Ltd, Glasgow. 2,100 I.H.P — 11 kts.
11.11.1939 launched and 1.1940 completed by Burntisland Sbg Co Ltd, Burntisland. 9.5.1941: damaged by bomb at Hull. 1946: Cook Sg Co Ltd, Jersey (Ambrose, Davies & Matthews Ltd, Swansea). 1949: Jersey United Sg Co Ltd, Jersey. 1952: United Transports Ltd, Jersey. 1960: Transoceanic Sg Co (World Wide Co (Ship Managers) Ltd), Hong Kong, renamed OCEANIC GEM. 1961: Gladiator Sg Co Ltd (World Wide (Shipping) Ltd), Hong Kong, renamed TOSA BAY. 1.2.1967: breaking up by Wise Investment Co commenced at Hong Kong.

DAN-Y-BRYN. *(Skyfotos)*

**GER-Y-BRYN** Price/Loan: £123,000/£110,000
O.N. 168173. Details as DAN-Y-BRYN, except 5,108g. 3,015n.
14.3.1941 launched and 5.1941 completed by Burntisland Sbg Co Ltd, Burntisland. 5.3.1943: torpedoed by U130 in 43.50N 14.45W, convoy XK2. Lagos for Hull, general. 46. crew, all saved. Grant paid on both vessels.

**Asiatic S.N. Co Ltd, London**
Grant paid on HAVILDAR (5,407/40) and RISALDAR (5,407/40).

**Australind S.Sg. Co Ltd, London**
Loan agreed on vessel by Denny (1,347). Price/Loan: £169,850/£120,000. Loan not taken up. Grant paid. This was the ARDENVOHR (5,025/40).

**B. & S. Sg Co Ltd, Cardiff**
**South American Saint Line Ltd** (prior to 4.4.1939 named The Barry Sg Co Ltd)
**ST ESSYLT** Price/Loan: £181,500/£125,000
O.N. 162150. 5,634g. 3,308n. 9,600d. 442.5 x 58.1 x 24.7 feet.
Doxford oil engine by Richardsons, Westgarth & Co Ltd, Hartlepool. 3,200 B.H.P. — 12 kts.
23.5.1940 launched and 9.1941 completed by Joseph L. Thompson & Sons Ltd, Sunderland. 4.7.1943: torpedoed by U375 in 36.44N 1.31E, 15 miles off Sicily, just prior to landing, convoy KMS18B. Clyde for Sicily, military stores. Caught fire, burnt all night and blew up at 5.45am 5.7.1943. 79 crew, 1 lost. Grant paid.

**J. A. Billmeir & Co Ltd, London**
**Stanhope S.S. Co Ltd**
**STANMORE** Price/Loan: £122,346/£97,500
O.N. 167623. 4,970g. 2,881n. 9,400d. 425.6 x 56.8 x 25.5 feet.
T 3-cyl by G. Clark (1938) Ltd, Sunderland. 1,750 I.H.P. — 10.5 kts.
22.5.1940 launched and 8.1940 completed by William Pickersgill & Sons Ltd, Sunderland. 1.10.1943: torpedoed by U223 in 36.41N 1.10E. 3.10.1953: towed in and beached at Tenes. 4.3.1944: decks buckling, C.T.L. Middlesbrough for Sicily, military stores. 49 crew, all saved. Grant paid.

Loan on vessel by Pickersgill (245) refused, as the order was placed on 4.10.1939 (one day after the scheme closed). This was the STANFORD (5,969/41).

Photographs taken by the late Lieutenant-Commander Jeffery W. Curtis, R.N.R., who as a R.N.V.R. Lieutenant, was in command of H.M. ML458. The **ST. ESSYLT** (above) seen on 10.11.1942 off "Z" Beach at Arzew during Operation "Torch", the Allied landings in French North Africa. The **STANMORE** (below) had sailed from Middlesbrough on 11.9.1943 for Sicily and is pictured after being torpedoed, in tow of H.M. Trawler FYLLA for beaching at Tenez. *(L. Dunn)*

**Bolton S.Sg. Co Ltd, London**

**RIBERA** Price/Loan: £117,507/£95,000
O.N. 167420. 5,559g. 3,337n. 9,700d. 432.2 x 58.0 x 26.8 feet.
T 3-cyl by D. Rowan & Co Ltd, Glasgow. 2,400 I.H.P — 11 kts.
22.2.1940 launched and 4.1940 completed by Lithgows Ltd, Port Glasgow. 1955: Cia Roblon de Oro S.A., Panama (Goulandris Bros Ltd, London), renamed OKEANIS. 1959: Cia de Nav Jolanda S.A., Monrovia (Soc Armamento Marittimo in Nome Collettivo [SOARMA], Genoa), renamed JOLANDA. 1964: Cia de Nav Victoria S.A., Panama (SOARMA, Genoa). 10.6.1971: delivered Split, breaking up by Brodospas commenced 16.6.1971.

**REMBRANDT** Price/Loan: £119,957/£95,000
O.N. 168048. Details as RIBERA.
30.8.1940 launched and 1.1941 completed by Lithgows Ltd, Port Glasgow. 1957: Navigation Transport Co Inc, Monrovia (Loucas Nomicos, Piraeus), renamed CAPETAN ANTONIS. 1964: renamed MASTROMITSOS. 1965: Bluebird Maritime Co. Ltd, Monrovia (Nicholas D. Koulos, Piraeus). 29.4.1967: arrived Kaohsiung for breaking up.
Grant paid on RIBERA, but not REMBRANDT.

**REMBRANDT.** *(Skyfotos)*

**British Channel Islands Sg Co Ltd, London**
(Controlled by Coast Lines Ltd from 1942)
**CHANNEL QUEEN** Price/Loan: £27,758/£22,000
O.N. 167630. 567g. 275n. 700d. 169.8 x 28.1 x 9.7 feet.
Polar oil engine (aft) by British Auxiliaries Ltd, Glasgow. 800 B.H.P. — 10.5 kts.
10.6.1940 launched and 9.1940 completed by Burntisland Sbg Co Ltd, Burntisland. 1947: renamed CHANNEL COAST. 1958: Zillah Sg Co Ltd (Coast Lines Ltd), Liverpool, renamed GLENFIELD. 1960: British Channel Islands Sg Co Ltd (Coast Lines Ltd), London, renamed ALDERNEY COAST. 1966: Nicolaos Grigoriou and others (George M. Moundreas & Bros), Piraeus, renamed ASTRONAFTIS. 1975: renamed SEA HORSE. 1975: Namar Sg Ltd (Matsinos Bros), Piraeus, renamed MASTRO COSTAS. 21.2.1976: engine breakdown, Greece for Lagos, pipes and tomato paste. 23.2.1976 towed into Monrovia by RETHYMNON. 10.3.1978: towed out of port by National Port Authority as now abandoned and a hazard to the port. Beached between Monrovia and Lofa River.

**CHANNEL QUEEN** on trials. Note protection of wheelhouse. *(Henry Robb Ltd.)*

**TUDOR QUEEN.** *(E. N. Taylor)*

**CORAL QUEEN** Price/Loan: £15,363/£12,000
O.N. 168090. 303g. 148n. 380d. 132.0 x 24.6 x 7.4 feet.
Oil engine (aft) Crossley Bros Ltd, Manchester. 330 B.H.P. — 9.25 kts.
25.6.1940 launched and 1.1941 completed by Burntisland Sbg Co Ltd, Burntisland. 1948: renamed CORAL COAST. 1949: Government of Gambia, Bathurst, renamed FULLADU. 9.1.1965: towed from Bathurst to Las Palmas for breaking up.

**TUDOR QUEEN** Price/Loan: £36,670/£29,000
O.N. 168071. 1,029g. 582n. 1,400d. 204.2 x 32.8 x 13.2 feet.
T 3-cyl (aft) by D. Rowan & Co Ltd, Glasgow. 600 I.H.P. — 10.25 kts.
31.12.1940 launched and 2.1941 completed by Burntisland Sbg Co Ltd, Burntisland. 1946: British Channel Traders Ltd (Coast Lines Ltd), London. 1947: Queenship Nav Ltd (Coast Lines Ltd), London. 1959: Coast Lines Ltd, London. 1959: broken up in U.K.
No grants paid.
Note: The TUDOR QUEEN was the prototype for a class of 10 Empire ships.

**British India S.N. Co Ltd, London**
Grant paid on OZARDA (6,895/40), ITRIA (6,845/40), ITOLA (6,793/40), ITAURA (6,793/40), ISMAILA (6,793/40) and IKAUNA (6,793/41).

**Buries Markes Ltd, London**
**LA ESTANCIA** Price/Loan: £155,000/£120,000
O.N. 167381. 5,185g. 3,050n. 9,340d. 429.8 x 56.5 x 26.5 feet.
Oil engine by builders. 3,300 B.H.P — 13 kts.
14.9.1939 launched and 1.1940 completed by Wm. Doxford & Sons Ltd, Sunderland. 19.10.1940: torpedoed by U47 in 57N 17W, convoy HX79. Mackay for Methil & Tees, sugar. 33 crew and 1 passenger, 1 crew missing.

**LA CORDILLERA** Price/Loan: £155,000/£120,000
O.N. 167411. Details as LA ESTANCIA.
28.11.1939 launched and 3.1940 completed by Wm. Doxford & Sons Ltd, Sunderland. 5.11.1942: torpedoed by U163 in 12.2N 58.4W. Suez for New York, ballast. 41 crew, 3 lost.
Grant paid on both vessels.

**Burnett S.S. Co Ltd, Newcastle**
Loan agreed on vessel by Laing (730). Price/Loan: £85,500/£60,000. Application withdrawn. Grant paid. This was the TYNEMOUTH (3,168/40). Loan application incorrectly gives the name TOWNELEY.

**Capper, Alexander & Co, London**
**Alexander Sg Co Ltd**
Grant paid on CHARLBURY (4,836/40).

**R. P. Care & Co Ltd, Cardiff**
Loan application refused for vessel by Lytham. This was stated to be a vessel to be launched in 1940, of 657g, 800d, and may have been similar to Zillah's BROOMFIELD (841) (657/38).

**James Chambers & Co, Liverpool**
**Lancashire Sg Co Ltd**
Grant Paid on BOLTON CASTLE (5,203/39).

NEWBROUGH. *(A. Duncan)*

**R. Chapman & Sons, Newcastle**
**Carlton S.S. Co Ltd and Cambay S.S. Co Ltd**
Grant paid on HERMISTON (4,813/39) and SCORTON (4,813/39).

**Charlton, McAllum & Co Ltd, Newcastle**
**Charlton S.S. Co Ltd**
Grant paid on HAZELSIDE (5,297/40).

**Claymore Sg Co Ltd, Cardiff**
**DAYDAWN** Price/Loan: £111,250/£75,000
O.N. 167798. 4,768g. 2,772n. 8,650d. 406.0 x 54.8 x 25.6 feet.
T 3-cyl by Richardsons, Westgarth & Co Ltd, Hartlepool. 1,450 I.H.P. — 10.25 kts.
9.12.1939 launched and 1.1940 completed by William Pickersgill & Sons Ltd, Sunderland. 21.11.1940: torpedoed by U103 in 56.30N 14.10W, convoy OB244. Barry for Rio Santiago, coal. 38 crew, 2 missing.
Grant paid.

**Common Brothers Ltd, Newcastle**
**Hindustan S.Sg. Co Ltd**
Loan agreed on vessel by Short (460). Price/Loan: £140,924/£110,000. Loan not required. Grant paid. This was the HINDUSTAN (5,245/40).

**Northumbrian Sg Co Ltd**
**NEWBROUGH** Price/Loan: £142,424/£110,000
O.N. 165807. 5,255g. 3,025n. 8,870d. 437.0 x 56.5 x 25.5 feet.
Oil engine by builders. 2,500 B.H.P. — 12 kts.
31.10.1940 launched and 2.1941 completed by Short Bros Ltd, Sunderland. 1955: Aviation & Shipping Co Ltd (Purvis Sg Co Ltd), London, renamed AVISBAY. 1962: Aegean Nav Co Inc, Monrovia (Loucas Nomicos, Piraeus), renamed GIANNIS. 10.1969: breaking up at Hong Kong completed by Fuji Marden & Co Ltd.
Grant paid.

**Halford Constant Ltd, London**
**Constants (South Wales) Ltd**
Loan agreed on vessel by Austin (355). Price/Loan: £60,000/£50,000. Loan not taken up. Grant paid. This was the OTTINGE (2,870/40).
Notes: Austin (355) was Gas Light and Coke's CAPITOL (1,558/41), OTTINGE was Gray (1,098). The Loan Committee noted the evasive behaviour of Constants (South Wales) Ltd in respect of the CAPE ORTEGAL under the Scrap and Build scheme. Bought from Constants by E.R. Management and credited against their Scrap and Build loan she was sold back to Constants (South Wales) Ltd for breaking up, but with no date in the contract by which she was to be scrapped. They kept her trading until 1939. Despite this a loan was agreed.

CAPE ORTEGAL. *(World Ship Photo Library)*

**Mitchell Cotts & Co Ltd, London**
**Saint Line Ltd**
Grant paid on SAINT BERNARD (5,183/39).

**Counties Ship Management Co Ltd, London**
**Dorset S.S. Co Ltd**
**PENTRIDGE HILL** Price/Loan: £133,549/£65,000
O.N. 168058. 7,579g. 5,589n. 9.880d. 421.2 x 60.5 x 35.8 feet. 'Arcform.'
T 3-cyl 'Reheater' engine by North Eastern Marine Engineering Co (1938) Ltd, Newcastle. 2,150 I.H.P. — 11 kts.
4.10.1940 launched and 1.1941 completed by Bartram & Sons Ltd, Sunderland. 1949: London & Overseas Freighters Ltd, London, renamed LONDON DEALER. 1951: Sociedad Transoceancia Canopus S.A., Panama (Rethymnis & Kulukundis Ltd, London), renamed CENTAURUS. 1961: Cia Naviera Adriatica Ltda, Beirut (Dabinovic S.A., Geneva), renamed NAJLA. 1965: broken up in Belgium by Jos Boel et Fils, Tamise.

PENTRIDGE HILL. *(A. Duncan)*

**LULWORTH HILL** Price/Loan: £127,050/£65,000
O.N. 167631. 7,628g. 5,595n. 9,900d. 421.1 x 60.4 x 35.8 feet. 'Arcform.'
T 3-cyl by D. Rowan & Co Ltd, Glasgow. 2,150 I.H.P. — 11 kts.
24.6.1940 launched and 9.1940 completed by William Hamilton & Co. Ltd, Port Glasgow. 19.3.1943: torpedoed by Italian submarine DA VINCI in 10.10S 1E. Mauritius for Liverpool, sugar and rum. 39 crew, 36 lost, 1 p.o.w., 2 survivors picked up from raft 7.5.1943.
Notes: Capital increased by £50,000 to satisfy the Committee requirements for granting of loan.

**Leith Hill Sg Co Ltd**
Application refused, but agreed on transfer to Putney Hill S.S. Co (KINGSTON HILL).

**Putney Hill S.S. Co Ltd**
**KINGSTON HILL** Price/Loan: £127,050/£65,000
O.N. 168045. Details as LULWORTH HILL.
17.10.1940 launched and 12.1940 completed by William Hamilton & Co Ltd, Port Glasgow. 22.2.1941: damaged by bomb from Condor of KG40 in 59.44N 12.33W, convoy OB287. Cardiff for Alexandria, coal, army vehicles and general. 40 crew, master died. 25.2.1941: arrived Loch Ewe in tow, repaired at Glasgow. 7.6.1941: torpedoed by U38 in 9.35N 29.40W. Cardiff and Glasgow for Alexandria, coal and general. 46 crew, 14 lost.

**PUTNEY HILL** Price/Loan: £140,000/£70,000
O.N. 167587. 5,216g. 3,066n. 9.490d. 427.6 x 56.5 x 26.5 feet.
Oil engine by builders. 2,500 B.H.P. — 12 kts.
21.3.1940 launched and 6.1940 completed by Wm. Doxford & Sons Ltd, Sunderland. 26.6.1942: sunk by torpedo and gunfire from U203 in 24.20N 63.16W. Haifa for New York, ballast. 38 crew, 3 lost.

**RICHMOND HILL** Price/Loan: £133,549/£65,000
O.N. 168042. Details as PENTRIDGE HILL, except engine built at Sunderland.
10.7.1940 launched and 11.1940 completed by Bartram & Sons Ltd, Sunderland. 1949: London & Overseas Freighters Ltd, London, renamed LONDON CRAFTSMAN. 1951: S.A. Importazione Carboni e Navigazione [SAICEN], Savona, renamed ITALGLORIA. 1951: renamed FIDUCIA. 1960: Cia de Nav Almirante S.A., Panama (Salvatores & C., S.R.L., Genoa), renamed SEARAVEN. 8.7.1966: arrived Yokosuka and 10.1966 breaking up completed by Amakusa Sangyo K.K.
Notes: Capital increased to £80,000 to satisfy Committee requirements for granting of loan.

**R. & K. & Tramp Sg (Cardiff) Ltd**
Application for loan refused, but agreed to transfer to Dorset S.S. Co (LULWORTH HILL).

**Tower S.S. Co Ltd**
TOWER GRANGE Price/Loan: £140,000/£85,000
O.N. 167617. Details as PUTNEY HILL, except 5,226g. 3,072n.
21.6.1940 launched and 7.1940 completed by Wm. Doxford & Sons Ltd, Sunderland. 18.11.1942: torpedoed by U154 in 6.20N 49.10W. Calcutta for U.K., ore and general. 47 crew, 6 lost.
Grant paid on PENTRIDGE HILL, PUTNEY HILL, RICHMOND HILL and TOWER GRANGE.
Notes: In 1948 the Dorset S.S. Co, Putney Hill S.S. Co and Tower S.S. Co were wound up and the assets transferred to a new company, London & Overseas Freighters Ltd, owned by the same Kulukundis and Mavroleon families.

**Craggs & Co, London**
Loan applications refused on two vessels by Burntisland (249 and 250). These two yard numbers were the EARLSTON (7,195/41) and ALLERTON (7,195/41) owned by the Carlton S.S. Co Ltd and Cambay S.S Co Ltd (R. Chapman & Son), Newcastle.

**Dalhousie Steam & Motor Ship Co Ltd, London**
Loan of vessel by Burntisland (230) refused. This was the DALHOUSIE (7,072/40). No grant paid.

**Dene Management Co Ltd, London**
**Elmdene Sg Co Ltd**
ELMDENE Price/Loan: £112,615/£112,615
O.N. 167371. 4,853g. 2,875n. 9,000d. 416.0 x 56.1 x 23.9 feet.
T 3-cyl by Central Marine Engine Works, West Hartlepool. 2,050 I.H.P. — 10.75 kts.
14.10.1939 launched and 12.1939 completed by William Gray & Co Ltd, West Hartlepool. 1940: Dene Sg Co Ltd (Dene Management Co Ltd), London. 8.6.1941: torpedoed by U103 in 8.16N 16.50W. Tyne for Alexandria, coal, service stores and aircraft. 36 crew, all saved.
Grant paid.

**J. & J. Denholm Ltd, Glasgow**
**The Denholm Line Steamers Ltd**
BROOMPARK Price/Loan: £115,000/£100,000
O.N. 166988. 5,136g. 3,057n. 9,200d. 431.9 x 56.2 x 24.8 feet.
T 3-cyl by D. Rowan & Co Ltd, Glasgow. 1,950 I.H.P. — 10.5 kts.
12.9.1939 launched and 10.1939 completed by Lithgows Ltd, Port Glasgow. 21.9.1940: damaged by torpedo from U48 in 55.8N 18.30W, convoy HX72. Vancouver for Glasgow, lumber & metal. 1 killed. Volunteer crew placed aboard. 23.9.1940: bombed off Islay. 25.9.1940: arrived Rothesay Bay, making water. 14.10.1940: arrived Greenock. 25.7.1942: torpedoed by U552 in 49.2N 40.26W. Taken in tow by CHEROKEE but sank 1.8.1942 in 47.42N 51.55W. Tyne for New York, ballast. 49 crew, 4 lost.

GLENPARK as STAD ARNHEM. *(Skyfotos)*

GLENPARK Price/Loan: £115,000/£100,000
O.N. 166939. Details as BROOMPARK except engines by Rankin & Blackmore Ltd, Greenock. 28.9.1939 launched and 11.1939 completed by Lithgows Ltd, Port Glasgow. 1951: Halcyon Lijn N.V., Rotterdam, renamed STAD ARNHEM. 1961: Marfomento Cia Nav S.A., Panama (Bray Sg Co Ltd. London), renamed ELLI. 1966: Maritime Coal Transport S.A.L.. Beirut (N. Guida, Naples), renamed FECONDO. 17.12.1970: aground in storm on Iles Cani, off Bizerta. Sicily for Algeria, pumice stone. 20.12.1970: refloated and towed Bizerta. Comp. T.L. Sold and breaking up commenced 7.1971 at La Spezia by Cantieri Navale Santa Maria S.p.A.

**Ellerman Lines Ltd, London**
**City Line Ltd**
Grant paid on CITY OF CALCUTTA (8,063/40).

**Ellerman & Papayanni Lines Ltd**
Grant paid on FLORIAN (3,174/40).

**Ellerman's Wilson Line Ltd, Hull**
Grant paid on VASCO (2,878/39), ANGELO (2,199/40) and ARIOSTO (2,176/40).

**William France, Fenwick & Co Ltd**
Grant paid on MOORWOOD (2,056/40) and CORNWOOD (2,777/40).

**Fenwick Fisher S.S. Co Ltd**
Loan on vessel by Austin (353) refused. Grant paid. This was the SEA FISHER (2,950/40).

**Furness, Withy & Co Ltd, London**
**Manchester Liners Ltd, Manchester**
Grant paid on MANCHESTER MERCHANT (7,264/40) and MANCHESTER TRADER (5,671/41).

**Prince Line Ltd**
**NORMAN PRINCE** Price/Loan: £83,000/£70,000
O.N. 167419. 1,913g. 919n. 3,400d. 304.0 × 44.2 × 16.5 feet.
T 3-cyl by builders. 1,900 I.H.P. — 12.75 kts.
23.12.1939 launched and 4.1940 completed by Smith's Dock Co Ltd, Middlesbrough. 28.5.1942: torpedoed by U156 in 14.40N 62.15W. Liverpool and Barranquilla for St Lucia, ballast. 40 crew and nine passengers, 14 crew and two passengers lost.

**LANCASTRIAN PRINCE** Price/Loan: £83,000/£70,000
O.N. 167486. Details as NORMAN PRINCE, except 1,914g. 920n.
7.3.1939 launched and 5.1940 completed by Smith's Dock Co Ltd, Middlesbrough. 11.4.1943: torpedoed by U613 in 50.18N 42.48W, convoy ON176. Partington and Mersey for Boston and/or St John NB. 38 crew and seven passengers, no survivors.

**WELSH PRINCE** Price/Loan: £123,000/£104,000
O.N. 167597. 5,148g. 3,053n. 8,700d. 418.1 × 56.0 × 25.0 feet.
T 3-cyl by D. Rowan & Co Ltd, Glasgow. 2,100 I.H.P. — 10.5 kts.
23.4.1940: launched and 6.1940 completed by Blythswood Sbg Co Ltd, Glasgow. 7.12.1941: mined 110° 5 cables from No 59 Buoy, Spurn Head. Abandoned in 53.24N 0.59E, aground in 53.23.40N 0.58.55E, broke in two. London for New York, general. 41 crew and 6 passengers, all saved.

TUDOR PRINCE. *(World Ship Photo Library)*

**TUDOR PRINCE** Price/Loan: £83,000/£70,000
O.N. 167616. Details as NORMAN PRINCE, except 1,914g. 920n.
23.5.1939 launched and 8.1940 completed by Smith's Dock Co Ltd, Middlesbrough. 1957: F. Italo Croce S.p.A., Genoa, renamed CROCE ITALO. 1961: Maritime Enterprise Co, Monrovia (S. Scalisi, Genoa), renamed ORNELLA. 1.7.1964: breaking up commenced by Cantieri Tommaso di Savoia, La Spezia.

**STUART PRINCE** Price/Loan: £83,000/£70,000
O.N. 168022. Details as NORMAN PRINCE, except 1,91 ,g.
15.8.1940 launched and 10.1940 completed by Smith's Dock Co Ltd, Middlesbrough. 1951: renamed FORT HAMILTON. 1958: renamed STUART PRINCE. 1959: Cia Mar Med Ltda, San Jose (D. Th. Petropoulos, London), renamed HALCYON MED. 24.8.1960: collision with tanker ESSO SWITZERLAND (23,363/59) in thick fog 120 miles E of Gibraltar. Cut in two, afterpart sank in 36.9N 3.36W. Forepart taken in two, sank 25.8.1960 in 36.26N 3.20W. Arzew for Granton, esparto grass.
Grant paid on all five vessels.

**Hain S.S. Co Ltd, London**
Grant paid on TREVETHOE (5,257/40), TREVAYLOR (5,257/40) and TREVILLEY (5,300/40).

**Haldin & Phillips Ltd, London**
**Court Line Ltd**
Grant paid on LAVINGTON COURT (5,372/40).

**Hall Brothers, Newcastle**
**Hall Bros S.S. Co Ltd**
Grant paid on ROYAL EMBLEM (4,900/40).

**J. & C. Harrison Ltd, London**
**Gowland S.S. Co Ltd**
Grant paid on HARPAGUS (5,173/40) and HARPALYCE (5,280/40).

**T. & J. Harrison, Liverpool**
**Charente S.S. Co Ltd**
Grant paid on TRADER (6,087/40), NOVELIST (6,183/40) and DALESMAN (6,343/40).

**Headlam & Son, Whitby**
**Rowland & Marwood's S.S. Co Ltd**
Grant paid on BARNBY (4,813/40).

**P. Henderson & Co, Glasgow**
**British & Burmese S.N. Co Ltd**
Grant paid on KALEWA (4,389/40).

**G. Heyn & Sons Ltd, Belfast**
**Ulster S.S. Co Ltd**
Grant paid on FANAD HEAD (5,038/40).

**H. Hogarth & Sons, Glasgow**
**Hogarth Sg Co Ltd**
Grant paid on BARON SCOTT (4,574/40).

**Kelvin Sg Co Ltd**
Grant paid on BARON HERRIES (4,574/40).

**John I. Jacobs & Co Ltd, London**
**BEECHWOOD** Price/Loan: £117,000/£100,000
O.N. 167385. 4,897g. 2,757n. 9,000d. 415.1 × 58.2 × 24.8 feet.
T 3-cyl 'Reheater' by North Eastern Marine Engineering Co (1938) Ltd, Newcastle. 1,450 I.H.P. — 10.5 kts.
9.11.1939 launched and 1.1940 completed by Sir James Laing & Sons Ltd, Sunderland. 3.2.1940: damaged by aircraft bombs and gunfire 3 miles E of Smith's Knoll Light Vessel. Tyne for Gibraltar, coal. Kept afloat by pumps, diverted to Purfleet to discharge. 26.8.1942: torpedoed by U130 in 5.30N 14.4W. Haifa and Lourenco Marques for U.K., general and potash. 41 crew and one stowaway, one killed, master p.o.w.
Grant paid. Grant also paid on GLENWOOD (4,897/40).

**Richard W. Jones & Co, Newport**
**Uskside S.S. Co Ltd**
Loan agreed on vessel by Burntisland (236). Price/Loan: £66,861/£60,000. Loan not taken up. Grant paid. This was the USKBRIDGE (2,700/40).

**Kaye, Son & Co Ltd, London**
**'K' S.S. Co Ltd**
Grant paid on MARSDALE (4,890/40).

**Lambert Brothers Ltd, London**
**Temple S.S. Co Ltd**
**TEMPLE ARCH** Price/Loan: £115,000/£115,000
O.N. 167384. 5,138g. 3,039n. 9,200d. 431.9 × 56.2 × 24.8 feet.
T 3-cyl by Rankin & Blackmore Ltd, Greenock. 2,350 I.H.P. — 11 kts.

**TEMPLE INN.** *(University of Glasgow)*

27.11.1939 launched and 1.1940 completed by Lithgows Ltd, Port Glasgow. 1959: Marivic Nav Inc, Monrovia (A. J. & D. J. Chandris, Piraeus), renamed MARIHORA. 1964: General Carriers S.A., Panama (A. J. & D. J. Chandris, Piraeus). 3.11.1966: wrecked in cyclone 2 miles S of Madras. Singapore for Madras, ballast.

**TEMPLE INN** Price/Loan: £119,370/£119,000
O.N. 167398. Details as TEMPLE ARCH, except 5,218g. 3,085n.
27.12.1939 launched and 2.1940 completed by Lithgows Ltd, Port Glasgow. 1959: Marivic Nav Inc, Monrovia (A. J. & D. J. Chandris, Piraeus), renamed MARIVIKI. 1964: General Carriers S.A., Panama (A. J. & D. J. Chandris, Piraeus). 21.7.1965: sprang a leak, beached near Mormugao. Madras for Constantza, iron ore. Broke in two. T.L.
Grant paid on both vessels.

**Lamport & Holt Line Ltd, Liverpool**
Grant paid on DEFOE (6,245/40) and DEBRETT (6,244/40).

**DUKE OF ATHENS.** *(A. Duncan)*

**S. Livanos & Co Ltd, London**
**Trent Maritime Co Ltd**
**DUKE OF ATHENS** Price/Loan: £138,500/£60,000
O.N. 168011. 5,217g. 3,025n. 9,500d. 427.0 × 56.5 × 26.5 feet.
Oil engine by builders. 2,500 B.H.P. — 12 kts.
18.7.1940 launched and 10.1940 completed by Wm. Doxford & Sons Ltd, Sunderland. 1961: Atlantic Freighters Ltd, Monrovia (S. Livanos, Chios), renamed BREEZE. 1965: Cia Nav Prodromos S.A., Panama (M. J. Lemos & Co Ltd, London), renamed SAN JOHN P. 1967: Mardinamico Cia Nav S.A., Panama (Manlemos Sg Agencies Co Ltd, Piraeus), renamed THEOKLETOS. 21.9.1969: arrived Karachi, broken up by Sind Steel Co.

**DUKE OF SPARTA** Price/Loan: £132,500/£60,000
O.N. 168027. 5,397g. 3,161n. 9,000d. 441.1 × 57.8 × 25.4 feet.
T 3-cyl by Central Marine Engine Works, West Hartlepool. 2,600 I.H.P. — 11.25 kts.

DUKE OF SPARTA. *(A. Duncan)*

9.7.1940 launched and 10.1940 completed by William Gray & Co Ltd, West Hartlepool. 1951: Fratelli Grimaldi, Naples, renamed AQUILA. 28.4.1958: bombed by Indonesian rebel aircraft at Amboina, 1.5.1958: bombed again, sank 27.5.1958. Waiting to load copra for Copenhagen.
Grant paid on both vessels.
Notes: Capital increased to £120,000.

**Comben Longstaff & Co Ltd, London**
**Williamstown Sg Co Ltd**
WORTHTOWN Price/Loan: £29,170/£15,000
O.N. 167361. 868g. 457n. 1,200d. 197.7 × 30.7 × 12.1 feet.
T 3-cyl (aft) by builders. 775 I.H.P. — 10 kts.
30.9.1939 launched and 11.1939 completed by John Lewis & Sons Ltd, Aberdeen. 27.5.1940: bombed, set on fire, sunk at Dunkirk during evacuation. 11.1940: raised, repaired (declared prize by Hamburg Prize Court 13.10.1943). Atlas Reederei A.G. (Schulte & Bruns), Emden, renamed ILSE SCHULTE. 5.1945: found in Schlei with bomb damage, recovered. 1945: Ministry of Transport (Comben Longstaff & Co Ltd), London, renamed EMPIRE WORTHTOWN. 1946: Williamstown Sg Co Ltd, renamed GLAMORGANBROOK. 11.10.1946: sprang a leak, sank 5 miles E of Scarborough. Blyth for Cowes, coal.
Grant paid.

**Lyle Sg Co Ltd, Glasgow**
**Cape of Good Hope Motorship Co Ltd**
Grant paid on CAPE RODNEY (4,512/40) and CAPE WRATH (4,512/40).

**J. D. McLaren & Co, London**
**Gowan Sg Co Ltd**
Loan on two vessels by Burntisland (233 and 234) refused.
These yard numbers were the MERTON (7,195/41) of the Carlton S.S. Co Ltd and Cambay S.S. Co Ltd (R. Chapman & Son), Newcastle, and the AMBROSE FLEMING (1,555/41) of the London Power Co Ltd (Stephenson Clarke Ltd), London.

EDENCRAG. *(L. Dunn)*

**Magee, Son & Co, West Hartlepool**
**Hartlepools S.S. Co Ltd**
EDENCRAG Price/Loan: £49,663/£30,000
O.N. 160787. 1,592g. 937n. 2,740d. 264.3 × 40.3 × 16.2 feet.
T 3-cyl by Aitchison, Blair Ltd, Greenock. 950 I.H.P. — 10 kts.
3.9.1940 launched and 12.1940 completed by Burntisland Sbg Co Ltd, Burntisland. 14.12.1942: torpedoed by U443 in 35.49N 1.25W. Gibraltar for North Africa, military stores. 26 crew, 13 lost.
Grant paid.

**Metcalfe, Son & Co Ltd**
**Metcalfe Sg Co Ltd**
INDUSTRIA Price/Loan: £120,000/£75,000
O.N. 160782. 4,861g. 2,881n. 9,000d. 416.0 × 56.1 × 23.9 feet.
T 3-cyl by Central Marine Engine Works, West Hartlepool. 2,050 I.H.P. — 10.75 kts.
23.3.1940 launched and 5.1940 completed by William Gray & Co Ltd, West Hartlepool. 5.5.1941: damaged by aircraft bombing at Liverpool. 1957: Socoa Sg Co Ltd, Monrovia (Ramon de la Sota, Jr, Biarritz), renamed TERESA. 8.11.1968: arrived Hirao for breaking up by Matsukura Kaiji K.K.
Grant paid.

INDUSTRIA. *(World Ship Photo Library)*

**Morel Ltd, Cardiff**
**Nolisement S.S. Co Ltd**
Grant paid on BEIGNON (5,218/39).

**Pontypridd S.S. Co Ltd**
Grant paid on CATRINE (5,218/40).

**John Morrison & Co, Newcastle**
**Morrison S.S. Co Ltd**
ELMDALE Price/Loan: £119,916/£95,000
O.N. 165808. 4,872g. 2,810n. 9,225d. 424.2 × 57.0 × 25.8 feet.
T 3-cyl by Rankin & Blackmore Ltd, Greenock. 1,900 I.H.P. — 10.5 kts.
14.12.1940 launched and 4.1941 completed by Burntisland Sbg Co Ltd, Burntisland. 6.4.1942: damaged by gunfire from Japanese submarine I-3 in 6.52N 78.50E. Karachi for Durban. Put into Colombo for temporary repairs, permanent repairs in U.S.A. 1.11.1942: torpedoed by U174 in 0.17N 34.55W. Baltimore for Alexandria, coal, military stores and general. 42 crew, six lost.
Grant paid.

ELMDALE. *(Henry Robb Ltd.)*

**William B. Nisbet, Newcastle**
**Tanfield S.S. Co Ltd**
Grant paid on LEA GRANGE (2,969/39).

**James Nourse Ltd, London**
Grant paid on INDUS (5,187/40) and SUTLEJ (5,189/40).

**T. Bowen Rees & Co Ltd, Alexandria**
**New Egypt & Levant Sg Co Ltd**
**ANTAR** Price/Loan: £139,989/£112,000, but only £70,000 of loan taken up.
O.N. 168066. 5,222g. 3,034n. 9,450d. 427.0 × 56.5 × 26.5 feet.
Oil engine by builders. 2,500 B.H.P. — 12 kts.
1.11.1940 launched and 2.1941 completed by Wm. Doxford & Sons Ltd, Sunderland. 1948: British India S.N. Co Ltd, London, renamed GARBETA. 12.4.1963: arrived Hong Kong, 27.4.1963 breaking up commenced by Hong Kong Salvage & Towage Co Ltd.
Grant paid.

**ANTAR,** anchored off Halifax 21 July 1943. *(National Maritime Museum)*

**Rennie & Watson, Glasgow**
Loan refused on vessel by Lithgows (936).
This yard number was the AIRCREST (5,237/40) of Crest Sg Co Ltd (Overseas Navigation Trust Ltd), London.

**Hugh Roberts & Son, Newcastle**
**North Sg Co Ltd**
**NORTH BRITAIN** Price/Loan: £114,000/£85,000.
O.N. 165801. 4,635g. 2,683n. 8,300d. 405.6 × 53.6 × 24.4 feet.
T 3-cyl 'Reheater' by North Eastern Marine Engineering Co (1938) Ltd, Newcastle. 1,700 I.H.P. — 10.5 kts.
16.9.1940 launched and 11.1940 completed by J. Readhead & Sons Ltd, South Shields. 4.5.1943: torpedoed by U125 in 55.8N 42.43W, straggler from convoy ONS5. Sank in 5 minutes. Glasgow for Halifax, fireclay. 46 crew, 34 missing.
Grant paid.

**Sir R. Ropner & Co Ltd, West Hartlepool**
**Pool Sg Co Ltd**
Grant paid on FISHPOOL (4,950/40) and SEAPOOL (4,820/40).

**Ropner Sg Co Ltd**
Grant paid on WANDBY (4,947/40).

**Royal Mail Lines Ltd, London**
Grant paid on POTARO (5,410/40), PALMA (5,419/41), PAMPAS (5,400/40) and PARDO (5,400/40). The intended name for the PALMA was PELOTAS.

**Sir William Reardon Smith & Sons Ltd, Cardiff**
**Leeds Sg Co Ltd**
Grant paid on ATLANTIC CITY (5,188/41), EASTERN CITY (5,185/41) and ORIENT CITY (5,095/40).

**Reardon Smith Line Ltd**
Grant paid on MADRAS CITY (5,092/40).

**Stephens, Sutton Ltd, Newcastle**
**REAVELEY** Price/Loan: £126,500/£100,000
O.N. 165802. 4,998g. 3,008n. 9,240d. 423.2 × 54.2 × 26.1 feet.
Oil engine by builders. 1,800 B.H.P. — 11 kts.
24.8.1940 launched and 12.1940 completed by Wm. Doxford & Sons Ltd, Sunderland. 1948: Grenehurst Sg Co Ltd (Hadjilias & Co Ltd), London, renamed GRENEHURST. 1956: Buries Markes Ltd, London, renamed LA BARRANCA. 1959: Eastwind Nav Co Ltd (D. L. Wu), Hong

Kong, renamed WESTWIND. 1965: Dalcape Sg Co Ltd (International S.S. Co Ltd), Hong Kong, renamed UNIVERSAL MARINER. 17.11.1969: arrived Whampoa for breaking up.

**Red 'R' S.S. Co Ltd**

**RAWNSLEY** Price/Loan: £126,500/£100,000
O.N. 165786. Details as REAVELEY.
23.4.1940 launched and 7.1940 completed by Wm. Doxford & Sons Ltd, Sunderland. 8.5.1941: torpedoed by aircraft in 34.56N 26.13E. Haifa for Suda Bay, military stores. 9.5.1941 in tow for Makryallo Bay, Crete, to beach. Anchored Hierapetra Bay owing to bad weather, where set on fire and sunk 12.5.1941 by enemy aircraft — 34.59N 25.46.26E.

**Thomasson Sg Co Ltd**

**RODSLEY** Price/Loan: £125,000/£100,000
O.N. 165777. Details as REAVELEY, except 5,000g. 3,014n.
12.10.1939 launched and 12.1939 completed by Wm. Doxford & Sons Ltd, Sunderland. 1951: Rederi A/B Nordstjernan (A. A. Johnson), Stockholm, renamed RESERV. 1953: A/S Oddero (A. I. Langfeldt & Co.), Kristiansand, renamed SIRENES. 1963: Loutra Maritime Cpn, Panama (John Livanos & Sons Ltd, London), renamed MARCOS G.F. 1965: Benigno Lim, Manila, renamed SAMPAGUITA. 1966: Laguna Nav Co Inc, Manila, renamed PHILIPPINE SAMPAGUITA. 1972: broken up at Singapore.

**RODSLEY**. *(W. H. Brown)*

**ROOKLEY** Price/Loan: £126,500/£100,000
O.N. 165789. Details as REAVELEY.
18.5.1940 launched and 8.1940 completed by Wm. Doxford & Sons Ltd, Sunderland. 1956: Demetrios P. Morgaronis & Sons, Piraeus, renamed DESPOINA. 1966: Cristalinamar S.A., Panama (G. della Gatta, Naples), renamed JUMBO. 1970: renamed IBIS II. 24.6.1971: arrived Split, breaking up by Brodospas commenced 20.8.1971.
Grant paid on all four vessels.

**ROOKLEY**. *(World Ship Photo Library)*

**Stott, Mann & Fleming Ltd, Newcastle**
**Clive Sg Co Ltd**
**HOPETARN** Price/Loan: £165,000/£130,000
O.N. 165778. 5,231g. 3,135n. 9,770d. 418.9 × 57.4 × 25.6 feet.
Doxford oil engine by builders. 3,000 B.H.P. — 12.5 kts.
9.11.1939 launched and 1.1940 completed by Swan Hunter & Wigham Richardson Ltd, Wallsend-on-Tyne. 29.5.1943: torpedoed by U198 in 30.50S 39.32E. Calcutta and Colombo for Durban, Table Bay and U.K., general. 44 crew, seven lost, master p.o.w.
Grant paid.
Note: Capital increased by £30,000. Loan made to Clive Sg Co Ltd and not to a new proposed subsidiary.

**Frank C. Strick & Co Ltd, London**
**Strick Line (1923) Ltd**
Grant paid on AFGHANISTAN (6,992/40) and BALUCHISTAN (6,992/40).

**B. J. Sutherland & Co Ltd, Newcastle**
Grant paid on ARGYLL (4,897/39), SUTHERLAND (5,172/40) and INVERNESS (4,897/40)

**E. J. Sutton & Co Ltd, Newcastle**
**Confield S.S. Co Ltd**
Loan arranged on vessel by Thompson (597). Price/Loan: £117,000/£90,000. Not taken up. Grant paid. This was the CONFIELD (4,956/40).

**W. J. Tatem Ltd, Cardiff**
Grant paid on WINKLEIGH (5,468/40).

**Stanley & John Thompson Ltd, London**
**Silver Line Ltd**
Applied for loan for vessel by Thompson. Loan application withdrawn.

**Turnbull, Scott & Co, London**
**Turnbull, Scott Sg Co Ltd**
Loan agreed on vessel by Burntisland (237). Price/Loan: £137,774/£120,000. Loan not required. Grant paid. This was the EASTGATE (5,032/40).

**Watts, Watts & Co Ltd, London**
**Britain S.S. Co Ltd**
**TOTTENHAM** Price/Loan: £132,100/£125,000
O.N. 167541. 4,762g. 2,663n. 9,200d. 417.6 × 56.8 × 25.0 feet.
T 3-cyl 'Reheater' by North Eastern Marine Engineering Co (1938) Ltd, Newcastle. 1,600 I.H.P. — 10 kts.
21.3.1940 launched and 6.1940 completed by Caledon Sbg & Engineering Co Ltd, Dundee. 19.3.1941: damaged by mine at Southend Anchorage, 20.3.1941 arrived Gravesend in tow. London for Tyne and Halifax, ballast. 17.6.1941: captured by raider ATLANTIS (masquerading as TAMESIS) in 7.38S 19.12W. Tyne for Suez, general, military transport and stores. 17 crew picked up by MAHRONDA after 11 days in boat, landed Trinidad, 26 crew p.o.w.

**TWICKENHAM** Price/Loan: £131,100/£125,000
O.N. 168028. Details as TOTTENHAM.
17.9.1940 launched and 11.1940 completed by Caledon Sbg & Engineering Co Ltd, Dundee. 15.7.1943: damaged by torpedo from U135 in 28.36N 13.18W, convoy OS51. Hull for Buenos Aires, coal. 50 crew. Taken in tow, later proceeded Dakar under own steam, arrived 31.7.1943. Discharged, temporary repairs, sailed for U.K. 21.10.1944. 1958: Great Eastern Sg Co Ltd, Bombay, renamed JAG MATA. 18.3.1963: arrived Bombay for breaking up by Great Eastern Sg Co Ltd.

**TEDDINGTON** on trials, fitted for war service. *(Henry Robb Ltd.)*

**TEDDINGTON** Price/Loan: £131,100/£125,000
O.N. 168080. Details as TOTTENHAM.
10.1.1941 launched and 3.1941 completed by Caledon Sbg & Engineering Co Ltd, Dundee. 17.9.1941: torpedoed by E-boat S-51 in 53.4N 1.34E. London for Durban and Calcutta, general and nickel. On fire, taken in tow, went ashore 18.9.1941 2.75 miles ESE from Cromer Pier, 1 mile WSW from Overstrand, in 52.55.48N 1.22.36E. Back broken, T.L., some cargo salvaged. 52 crew, all saved. 1.10.1941 near miss by bomb. 7.1954: to be dispersed by explosives.
Grant paid on all three vessels.

**Andrew Weir & Co, London**
**Bank Line Ltd**
Loan agreed on vessel by Readhead (518). Price/Loan: £142,890/£142,890. Loan not taken up. This was the THURSOBANK (5,575/40).
Loan applied for on two vessels by Harland & Wolff, Belfast (1034 and 1035). This loan was refused as ineligible, the terms of the scheme not applying to Northern Ireland which still had its own Loan Guarantee Act in operation. These were the ARAYBANK (7,258/40) and SHIRRABANK (7,274/40).
Grant paid on all three vessels.

**West Hartlepool S.N. Co Ltd, West Hartlepool**
Applied for loan on vessel by Doxford (657), but in view of the extremely strong financial position of the firm the Committee felt a loan should not be made. Following informal talks with the Company the application was withdrawn. This was a "feeler" by the Company who had a future large building programme in mind. This was the DERWENTHALL (4,934/40). Grant paid on the DERWENTHALL, but refused for the DALTONHALL (5,175/41).

GRAIGLAS. *(G. E. P. Brownell)*

**Idwal Williams & Co, Cardiff**
**Graig Sg Co Ltd**
**GRAIGLAS** Price/Loan: £102,500/£60,000
O.N. 167801. 4,312g. 2,549n. 7,500d. 389.3 × 54.7 × 24.2 feet.
T 3-cyl 'Reheater' by G. Clark (1938) Ltd, Sunderland. 1,300 I.H.P. — 10 kts.
23.2.1940 launched and 5.1940 completed by Joseph L. Thompson & Sons Ltd, Sunderland. 1952: Lantao S.S. Co Ltd (Wallem & Co Ltd), Hong Kong, renamed LANTAO. 1966: General S.S. Co Ltd S.A., Hong Kong, renamed SHIA. 13.2.1967: arrived Hong Kong, breaking up commenced 11.3.1967 by Leung Yau Sbkg Co.
Grant paid. A loan application for a second vessel by Thompson was ineligible.

# Thirty-five Years of Cargo-Boat Earnings.

| Period. | | Paid-up capital. | Debentures, loans, etc. | Book value of fleet.† | Profit on voyages, etc. | Dividend. | | Fleet. | | Deprecia-tion written off. | Deprecia-tion at 5 per cent. |
|---|---|---|---|---|---|---|---|---|---|---|---|
| | | | | | | Amount. | Per cent. | No. | Tons gross. | | |
| | | £ | £ | £ | £ | £ | | | | £ | £ |
| Five year averages— | | | | | | | | | | | |
| 1904–1908 | ... | 8,608,632 | 3,892,040 | 12,661,860 | 921,486 | 321,764 | 3.74 | 463 | 1,419,093 | 322,079 | 760,896 |
| 1909–1913 | ... | 10,076,439 | 5,668,818 | 15,480,886 | 2,267,483 | 573,160 | 5.69 | 545 | 1,852,934 | 1,158,752 | 945,972 |
| 1914–1918 | ... | 13,227,492 | 8,775,012 | 19,608,484 | 5,581,064 | 1,929,667 | 14.59 | 502 | 1,912,136 | 2,143,198 | 967,377 |
| 1919–1923 | ... | 22,562,176 | 12,471 974 | 36,863,451 | 3,815,048 | 1,779,972 | 7.89 | 361 | 1,300,883 | 1,298,995 | 1,593,858 |
| Years— | | | | | | | | | | | |
| 1924 ... | ... | 27,694,369 | 9,772 699 | 41,906,917 | 2,211,919 | 1,049,395 | 3.78 | 413 | 1,786,697 | 1,159,039 | 1,944,199 |
| 1925 ... | ... | 26,397,029 | 10,481,077 | 42,308,657 | 2,211,683 | 996,664 | 3.77 | 445 | 1,933,110 | 1,050,213 | 1,974,047 |
| 1926 ... | ... | 26,291,824 | 9,737,996 | 42,212,648 | 2,005,193 | 949,559 | 3.61 | 462 | 2,024,011 | 1,079,813 | 1,870,608 |
| 1927 ... | ... | 26,179,396 | 10,545,203 | 42,629,951 | 3,145,699 | 1,358,699 | 5.18 | 488 | 2,133,285 | 1,537,405 | 1,843,516 |
| 1928 ... | ... | 27,447,917 | 14,261,049 | 41,444,373 | 4,999,005 | 2,080,586 | 7.58 | 576 | 2,653,581 | 2,602,573 | 2,390,851 |
| 1929 ... | ... | 18,134,035 | 8,564,998 | 29,226,189 | 1,466,690 | 928,391 | 5.12 | 489 | 2,070,359 | 605,487 | 1,525,013 |
| 1930 ... | ... | 16,748,121 | 8,400,577 | 24,478,792 | 1,363,064 | 652,398 | 3.89 | 454 | 2,075,038 | 537,215 | 1,542,367 |
| 1931 ... | ... | 15,722,698 | 7,712,101 | 22,418,769 | 607,023 | 241,438 | 1.53 | 431 | 2,026,890 | 531,834 | 1,472,126 |
| 1932 ... | ... | 14,585,461 | 5,887,248 | 19,611,259 | 626,325 | 229,905 | 1.58 | 387 | 1,850,556 | 551,887 | 1,226,534 |
| 1933 ... | ... | 14,001,163 | 5,331,014 | 18,330,350 | 155,023 | 135,380 | 0.97 | 361 | 1,733,989 | 349,598 | 1,193,389 |
| 1934 ... | ... | 14,636,389 | 5,811,210 | 19,223,343 | 88,653 | 208,246 | 1.42 | 347 | 1,701,940 | 134,304 | 1,190,607 |
| 1935 ... | ... | 12,189,426 | 4,165,392 | 16,489,400 | 499 235 | 178,545 | 1.46 | 316 | 1,495,455 | 342,939 | 1,026,291 |
| 1936 ... | ... | 12,451,926 | 3,914,[illegible]98 | 16,177,984 | 1,082,889 | 260,435 | 2.09 | 309 | 1,464,487 | 673,578 | 1,005,275 |
| 1937 ... | ... | 11,462,193 | 3,707,252 | 14,522,118 | 2,237,009 | 481,733 | 4.20 | 294 | 1,404,959 | 1,071,893 | 954,583 |
| 1938 ... | ... | 12,178,135 | 2,860,526 | 14,717,444 | 4,546,688 | 1,123,207 | 9.22 | 329 | 1 664,945 | 2,510,317 | 1,005,488 |

† Including, in some instances, investments

(Fairplay)

# Working of Some Cargo-boat Companies in 1934.

| Name of Company. | Paid-up Capital. | Debentures, Loans, Bills payable, sundry creditors. | Book Value of Steamers. | Sundry debtors, bills receivable, investments and cash. | Vessels. | | Profit or loss from Voyages, etc. | Dividend on Capital. | | Transferred to Depreciation out of earnings. | Depreciation at 5 per cent (a) |
|---|---|---|---|---|---|---|---|---|---|---|---|
| | | | | | No. | Tons Gross. | | Amount. | Per cent. | | |
| | £ | £ | £ | £ | | | £ | £ | | £ | £ |
| Aberdeen Coal & Shipg. Co. | 60,000 | 8,460 | 16,000 | 101,394 | 2 | 2,179 | +5,341 | 5,625 | 9.38 | 1,000 | 1,020 |
| Britain S.S. Co. ... ... | 500,000 | 40,817 | 440,050 | 40,875 | 11 | 48,662 | +1,247 | *nil* | — | *nil* | 26,403 |
| British & Burmese Co. ... | 400,000 | 16,451 | 267,225 | 372,769 | 7 | 41,753 | (*l*) +7,362 | 20,000 | 5.00 | — | 16,033 |
| **British Oil Shipping Co....** | 250,000 | 211,847 | 354,295 | 20,700 | 2 | 22,407 | +15,970 | *nil* | — | (*e*) *nil* | 22,698 |
| Burmah S.S. Co. ... ... | 380,000 | 23,108 | 267,225 | 243,735 | 6 | 41,753 | (*l*) +5,881 | 19,000 | 5.00 | — | 16,033 |
| Burnett S.S. Co. ... ... | 200,000 | 70,999 | 267,135 | 9,012 | 4 | 12,029 | —21,864 | *nil* | — | *nil* | 18,445 |
| **Cairn Line ... ... ...** | 600,000 | 20,493 | 546,210 | 86,406 | 7 | 35,660 | — 9,593 | *nil* | — | *nil* | 32,772 |
| Claymore Shipping Co. ... | 100,000 | 17,408 | 107,000 | 4,612 | 2 | 7,711 | —2,087 | *nil* | — | *nil* | 6,420 |
| Eclipse Shipping Co. ... | 40,779 | 28,443 | 64,864 | 2,272 | 1 | 4,751 | — 516 | *nil* | — | *nil* | 3,459 |
| **Fairwater S. Co. ... ...** | 34,730 | 23,710 | 53,000 | 6,569 | 1 | 4,108 | —164 | *nil* | — | *nil* | 3,180 |
| **Glasgow Shipowners Co. ...** | 120,600 | 642 | 268,812 | 52,283 | 4 | 20,116 | — 2,147 | *nil* | — | *nil* | 13,440 |
| Glasgow United S. Co. ... | 70,000 | 19,085 | 103,688 | 6,450 | 3 | 15,156 | —1,947 | *nil* | — | (*m*) *nil* | 6,221 |
| Graig Shipping Co. ... | 100,000 | 58,294 | 111,352 | 6,859 | 2 | 7,380 | +3,290 | *nil* | — | *nil* | 6,681 |
| Granite City S.S. Co. ... | 20,648 | 2,083 | 6,000 | 27,384 | 1 | 1,023 | +411 | *nil* | — | *nil* | 852 |
| **Hadley Shipping Co. ...** | 75,000 | 48,514 | 141,750 | 9,408 | 3 | 15,019 | +34,499 | 5,625 | 7.50 | 20,000 | 10,592 |
| Hall Brothers S.S. Co. ... | 162,500 | 16,113 | 138,300 | 57,117 | 7 | 31,681 | —3,380 | 2,500 | 1.54 | *nil* | 8,289 |
| Hindustan S.S. Co. ... | 540,000 | 348,483 | 864,938 | 98,206 | 16 | 85,502 | —13,016 | *nil* | — | (*k*) *nil* | 57,171 |
| Hogarth Shipping Co. ... | 900,000 | 59,836 | (*j*) 1,396,314 | 14,488 | 19 | 79,743 | (*l*) +41,785 | 45,000 | 5.00 | (*i*) | 83,779 |
| Iberia Shipping Co. ... | 38,400 | 1,969 | (*j*) 38,774 | 5,067 | 2 | 6,465 | (*l*) +16,449 | 1,920 | 5.00 | (*i*) | 2,326 |
| " Induna " S.S. Co. ... | 25,500 | 7,301 | 25,500 | 26,576 | 3 | 13,958 | —4,119 | 510 | 2.00 | *nil* | 1,530 |
| Kelvin Shipping Co. ... | 348,800 | 30,972 | (*j*) 666,560 | 20,851 | 16 | 50,234 | (*l*) +10,872 | 10,464 | 3.00 | (*i*) | 39,993 |
| **King Line ... ... ...** | 500,000 | 562,802 | 926,500 | 286,770 | 17 | 86,152 | +30,435 | *nil* | — | (*f*) 7,534 | 58,525 |
| **Lancashire Shipping Co. ...** | 320,000 | 208,653 | 688,585 | 56,652 | 8 | 46,172 | (*b*) —8,826 | *nil* | — | (*c*) — | 45,215 |
| Leeds Shipping Co. ... | 300,000 | 139,820 | 414,000 | 33,980 | 6 | 25,863 | +8,787 | *nil* | — | 11,464 | 19,845 |
| Llangollen S.S. Co. ... | 40,000 | 3,251 | 40,000 | 9,373 | 1 | 5,056 | —947 | *nil* | — | *nil* | 2,400 |
| Maritime Shipping Co. ... | 22,500 | 51,993 | 63,591 | 4,205 | 1 | 5,218 | +1,073 | *nil* | — | 3,271 | 3,815 |
| Monarch S.S. Co. ... ... | 317,600 | 2,093 | 327,362 | 264,228 | 4 | 23,175 | —5,540 | 15,880 | 5.00 | *nil* | 19,641 |

(Fairplay)

| Name of Company. | Paid-up Capital. | Debentures, Loans, Bills payable, sundry creditors. | Book Value of Steamers. | Sundry debtors, bills receivable, investments and cash. | Vessels | | Profit or loss from Voyages, etc. | Dividend on Capital. | | Transferred to Depreciation out of earnings. | Depreciation at 5 per cent (a) |
|---|---|---|---|---|---|---|---|---|---|---|---|
| | | | | | No. | Tons Gross. | | Amount. | Per cent. | | |
| | £ | £ | £ | £ | | | £ | £ | | £ | £ |
| **Moor Line** ... ... ... | 423,000 | 1,165,298 | 1,343,058 | 170,500 | 21 | 99,155 | +1,332 | *nil* | — | *nil* | 80,583 |
| Neptune S.N. Co. ... ... | 437,534 | 120,772 | 108,865 | 458,457 | 2 | 13,354 | +2,158 | *nil* | — | *nil* | 6,532 |
| **Nitrate Producers' S.S. Co.** | 156,300 | 32,849 | 290,369 | 397,300 | 7 | 41,316 | +28,673 | 11,722 | 7.50 | (h)10,000 | 35,424 |
| Norfolk & N. American S. Shipping Co. ... ... | 800,000 | 831,028 | 928,529 | 653,011 | 6 | 39,913 | —50,815 | *nil* | — | *nil* | 55,711 |
| North of England S.S. Co. | 226,782 | 111,110 | 263,430 | 16,574 | 4 | 19,835 | —18,235 | *nil* | — | *nil* | 15,805 |
| **Pool Shipping Co.** ... | 937,500 | 2,375 | 1,071,362 | 139,122 | 29 | 140,190 | +8,081 | *nil* | — | (d) *nil* | 69,381 |
| Reardon Smith Line ... | 1,232,000 | 153,084 | 1,245,000 | 177,493 | 22 | 115,728 | +31,287 | *nil* | — | 63,535 | 73,260 |
| Ropner Shipping Co. ... | 822,106 | 16,252 | 719,837 | 133,289 | 19 | 91,971 | +17,700 | *nil* | — | 13,500 | 44,000 |
| Scottish Nav. Co. ... ... | 48,980 | 11,961 | 228,123 | 884 | 4 | 17,800 | —6,063 | *nil* | — | *nil* | 11,406 |
| **Shakespear Shipping Co.** ... | 100,000 | 66,772 | 152,000 | 14,730 | 6 | 27,116 | —6,039 | *nil* | — | *nil* | 15,665 |
| Sheaf S.S. Co. ... ... | 150,000 | 229,097 | 374,182 | 16,489 | 9 | 28,126 | +13,217 | *nil* | — | 4,000 | 22,451 |
| Stag Line ... ... ... | 205,603 | 7,249 | 216,987 | 1,525 | 6 | 20,469 | —7,100 | *nil* | — | *nil* | 13,019 |
| Tankers ... ... ... | 1,021,408 | 38,252 | 884,635 | 351,231 | 10 | 70,828 | —28,639 | *nil* | — | *nil* | 53,078 |
| **Tatem S.N. Co.** ... ... | 350,000 | 27,191 | 205,190 | 2,382,812 | 5 | 25,266 | (g) | 70,000 | 20 | (i)— | 12,311 |
| Thompson S. Shipping Co. | 130,002 | 21,054 | 196,759 | 35,284 | 4 | 16,767 | +3,571 | *nil* | — | *nil* | 11,805 |
| Turnbull Scott Shipping Co. | 37,317 | 269,240 | 472,227 | 22,121 | 9 | 42,200 | —1,308 | *nil* | — | *nil* | 28,333 |
| **United British S.S. Co.** ... | 1,050,000 | 646,507 | 1,691,010 | 76,343 | 25 | 138,650 | —2,001 | *nil* | — | *nil* | 101,460 |
| West Wales S.S. Co. ... | 40,800 | 37,479 | 226,750 | 5,684 | 3 | 14,330 | —6,422 | *nil* | — | *nil* | 13,605 |
| Totals ... ... ... | 14,636,389 | 5,811,210 | 19,223,343 | 6,921,090 | 347 | 1,701,940 | + 88,653 | 208,246 | 1.42 | 134,304 | 1,190,607 |

(*a*) Depreciation is taken at 5 per cent. on original cost, or 6 per cent. on written down value. (*b*) Including adjustment of average accounts, *less* interest on investments. (*c*) £70,000 was transferred from the reserve account and £15,000 from the insurance account to profit and loss account, of which £65,000 was written off the fleet. (*d*) £60,000 was transferred to profit and loss account from reserves (£62,962 of repairs were charged to reserve) and £85,000 transferred for depreciation. (*e*) £24,000 was written off by increasing the debit balance. (*f*) A further £41,392, profit on sale of investments, was transferred to depreciation. (*g*) Extent of loss on trading not disclosed. (*h*) A further £20,000 was transferred from reserve. (*i*) Amount written off not stated. (*j*) Including investments. (*k*) £87,924 transferred from reserve. (*l*) After providing for depreciation. (*m*) £37,928 was transferred from reserve for depreciation.

# BRITISH SHIPPING (ASSISTANCE) ACT, 1935

An act to make provision for the granting of financial assistance to the owners of ships registered in the United Kingdom in respect of tramp voyages carried out during the year nineteen hundred and thirty-five, and to persons qualified to be owners of British ships in respect of proposals for the improvement of merchant shipping fleets; to provide for the repeal of section eighteen of the Economy (Miscellaneous Provisions) Act, 1926; and for purposes connected with the matters aforesaid.

(26th February 1935)

BE it enacted by the King's most Excellent Majesty, by and with the advice and consent of the Lords Spiritual and Temporal, and Commons, in this present Parliament assembled, and by the authority of the same, as follows:-

## PART I

### SUBSIDY IN RESPECT OF TRAMP VOYAGES

1.—(1) For the purpose of helping the owners of vessels registered at ports in the United Kingdom to compete with foreign shipping in receipt of subsidies from foreign Governments, the Board of Trade may, subject to such directions as may be given by the Treasury, and upon recommendations made by an advisory committee (hereinafter referred to as the "Tramp Shipping Subsidy Committee") appointed for the purposes of this Part of this Act by the Board with the concurrence of the Treasury, pay to the owners of vessels eligible for subsidy under this Part of this Act subsidies in respect of tramp voyages or parts of tramp voyages carried out by such vessels in the year nineteen hundred and thirty-five:

Provided that no such subsidies shall be paid in respect of any voyage wholly between ports within the United Kingdom, Irish Free State, Isle of Man and Channel Islands.

(2) The vessels eligible for subsidy under this Part of this Act are vessels to which this Act applies, being vessels registered at ports in the United Kingdom, which have been British ships since the first day of January, nineteen hundred and thirty-four or, in the case of vessels completed after that date, since they were completed, so, however, that vessels completed after that date shall not be eligible for subsidy unless they were built in the United Kingdom.

(3) It shall be the duty of the Tramp Shipping Subsidy Committee to advise the Board generally as to the operation of this Part of this Act, and in particular as to the making of payments by way of subsidies under this Part of this Act and the terms and conditions upon which such payments should be made, and in considering any recommendations to be made to the Board of Trade under this Part of this Act the Committee shall have regard to the purpose for which the Board are empowered to grant subsidies under this Part of this Act and shall not recommend payment of a subsidy in respect of any tramp voyage if, in the opinion of the Committee, the voyage was undertaken without due regard to the necessity for co-operation between owners of British vessels in furthering that purpose.

(4) The Board of Trade may appoint secretaries to the Tramp Shipping Subsidy Committee, and the Committee may employ such officers and servants as the Committee may, with the consent of the Board and of the Treasury, determine; and there shall be paid as part of the expenses incurred by or on behalf of the Board of Trade under this Part of this Act to the secretaries and to any officers and servants so appointed such salaries and allowances as may be determined by the Board with the approval of the Treasury.

(5) The sums necessary for the payment of subsidies under this section and for the payment of any expenses incurred by or on behalf of the Board of Trade under this Part of this Act shall be defrayed out of moneys provided by Parliament and shall not exceed in the aggregate two million pounds.

## PART II

### ASSISTANCE TO SHIPOWNERS PROPOSING TO IMPROVE MERCHANT SHIPPING FLEETS

2.—If the Board of Trade, after consultation with an advisory committee (hereinafter referred to as the "Ships Replacement Committee") appointed for the purposes of this Part of this Act by the Board with the concurrence of the Treasury, are satisfied that the carrying out of any proposals for the demolition and for the building or modernization of vessels submitted to the Board by any person qualified under the Merchant Shipping Act, 1894, to own a British ship will promote the general interests of British shipping, and that the proposals comply with the requirements of this Part of this Act, the Board may approve the proposals and upon any such proposals being so approved the Board may, with the consent of the Treasury, and upon a recommendation made by the Ships Replacement Committee, make an advance to that person for the purpose of enabling him to build or modernize the vessels provided by the proposals.

3.—(1) No proposals submitted under this Part of this Act shall be approved by the Board of Trade unless the Board are satisfied, in particular, that all vessels to be demolished, built or modernized in pursuance of the proposals are vessels in respect of which assistance may be given under this Part of this Act and that the proposals contain provisions securing—

(*a*) that the demolition of vessels will be in the proportions of two gross tons to be demolished for every gross ton of the vessels to be built and of one gross ton to be demolished for every gross ton of the vessels to be modernized;

(*b*) that the vessels to be demolished in pursuance of the proposals—

(i) will be demolished within the United Kingdom, or with the consent of the Board of Trade outside the United Kingdom in accordance with such conditions as may be imposed by the Board;

(ii) will not, after the date upon which any advance is made under this Part of this Act or such later date as the Board may allow, be used for any trading voyage;

(*c*) that the vessels to be built or modernized in pursuance of the proposals—

(i) will be built or modernized in Great Britain and registered at a port in the United Kingdom; and

(ii) will, while any part of the principal of, or interest on, any advance made under this Part of this Act remains outstanding, neither be sold without the consent of the Board nor cease to be registered at a port within the United Kingdom.

(2) The vessels in respect of which assistance may be given under this Part of this Act are vessels to which this Act applies (not being vessels constructed or adapted for the carriage of more than twelve passengers) which are, were, or will be, as the case may be, employed in the carriage of commercial cargoes and not employed mainly in voyages between ports within the United Kingdom, Irish Free State, Isle of Man and Channel Islands, or in maintaining regular services between such ports and ports in the Continent of Europe between the River Elbe and Brest inclusive.

4.—(1) The advances made under this Part of this Act shall not exceed in the aggregate ten million pounds, and no such advance shall be made after the expiration of two years from the date of the passing of this Act.

(2) Any advance made under this Part of this Act shall be made upon such terms as may be determined by the Treasury, and without prejudice to the generality of this provision such terms shall include the following terms, that is to say:—

(*a*) that the advance shall be secured by a first mortgage on any vessels in respect of the building or modernization of which the advance is made;

(*b*) that the rate of interest on any advance made shall not exceed three per cent per annum;

(*c*) that any advance made shall be repayable within a period not exceeding twelve years.

(3) Such sums as may from time to time be required for the purposes of making any advances under this Part of this Act shall be charged on and issued out of the Consolidated Fund of the United Kingdom or the growing produce thereof.

(4) For the purpose of providing for the issue of such sums out of the Consolidated Fund or of any part of any such sum so issued, the Treasury may raise money in any manner in which they are authorised to raise money under and for the purpose of subsection (1) of section one of the War Loan Act, 1919, and any securities created and issued to raise money under this subsection shall for all purposes be deemed to have been created and issued under the said subsection (1).

(5) All sums received by way of interest on, or in repayment of, advances made under this Part of this Act shall be applied in such manner as the Treasury may direct to the redemption of debt.

(6) The Board of Trade shall, before the first day of October in every year, prepare an account in such form and in such manner as the Treasury may direct of the advances made under this Part of this Act, and the sums received by way of interest on, or in repayment of, such advances during the last preceding financial year and the Comptroller and Auditor General shall examine and certify the account and shall lay copies thereof, together with his report thereon, before both Houses of Parliament.

5.—This Part of this Act shall come into operation upon such date as may be appointed by order of the Board of Trade with the consent of the Treasury.

## PART III

## MISCELLANEOUS

6.—(1) The vessels to which this Act applies are all ships which are neither fishing vessels nor constructed or adapted for the carriage of liquid cargoes in bulk, nor so constructed or adapted that the space insulated for the carriage of special cargoes is in excess either of fifty thousand cubic feet or ten per cent of the total space available for cargo.

(2) In this Act the expression "tramp voyage" means a voyage in the course of which all the cargo is carried under charter party, but does not include any voyage during any part of which more than twelve passengers are carried.

(3) If, in connection with the operation of this Act, any question arises—

(*a*) whether a vessel is a vessel to which this Act applies; or

(*b*) whether a vessel is eligible for subsidy under Part I of this Act; or

(*c*) whether any voyage was a tramp voyage, or as to the extent of any such voyage, or whether any such voyage or part of such a voyage was carried out in the year nineteen hundred and thirty-five, or was a voyage wholly between ports within the United Kingdom, Irish Free State, Isle of Man, and Channel Islands; or

(*d*) whether any vessel is a vessel in respect of assistance may be given under Part II of this Act,

the question shall be decided by the Board of Trade after consulation with the appropriate advisory committee, and the decision of the Board shall be final.

(4) Anything required or authorised by or under this Act to be done by, to, or before the Board

of Trade may be done by, to, or before the President or any Secretary, Under Secretary or Assistant Secretary of the Board or any person authorised in that behalf by the President of the Board.

7.—Section eighteen of the Economy (Miscellaneous Provisions) Act, 1926 (which gives directions to the Board of Trade as to fixing the amount of the fees to be charged under the Merchant Shipping Acts, 1894 to 1923) is hereby repealed.

8.—This Act may be cited as the British Shipping (Assistance) Act, 1935.

(Quoted by permission of H.M.S.O.)

## Appendix d

## SHIPS REPLACEMENT COMMITTEE (1935)

Committee consisted of a chairman, naval architect, marine engineer, cargo shipowner, tramp shipowner and businessman. Commenced operations 18 March 1935. Loans terminated 25 February 1937. Report dated 15 March 1937.

**Sir Charles Alexander Innes, K.C.S.I., C.I.E. (Chairman)**

Born 27 October 1874, died 28 June 1959.

Joined Indian Civil Service in Madras 1898: Under Secretary to the Government of India 1907-10: Collector of Malabar 1911-15: Director of Industries and Controller of Munitions, Madras 1916-18: Indian Foodstuffs Commissioner 1919: Secretary to Government of India, Commerce Dept, 1920-1: Member, Governor-General's Council, India 1921-7: Governor of Burma 1927-32.

C.I.E. 1919: C.S.I. 1921: K.C.S.I. 1924.

**Frederick George Bryant, M.I.N.A.**

Born 11 April 1892, died 1966.

Trained with A. Goodwin-Hamilton & Adamson, consultant naval architects and marine engineers, Liverpool 1908: A.I.N.A. 1925: M.I.N.A. 1930: Member of Council of I.N.A. 1950-2.

**Sir George Perrin Christopher, Kt**

Born 1890, died 24 November 1977.

Director, P & O S.N. Co: Chairman, Hain S.S. Co Ltd: Chairman and Manging Director, Union-Castle Mail S.S. Co Ltd: President, Chamber of Shipping of the U.K. 1948-9 (Vice-President 1947-8: member of Council 1927 on): Joint Chairman, General Council of British Shipping 1947-9: Member, Council of British Shipping from 1927: Member, General Committee of Lloyd's Register of Shipping: Member, Executive Council, Shipping Federation: Liveryman, Worshipful Company of Shipwrights: Chairman, Tramp Shipping Administrative Committee and London General Shipowners Society 1939: Deputy Director, Commercial Services, Ministry of War Transport, 1939-41, Director 1941-5.

Knighted 1946.

**Albert Edward Laslett, I.S.O., M.I.N.A., M.I.Mar.E.**

Born London 1866. Died 3 January 1937.

Trained as engineer with John Jones & Sons, Liverpool: at sea for nine years, including seven years with White Star Line: E. H. Bushell, consulting engineers, Liverpool, 1893: Board of Trade Marine Department 1897, Senior Surveyor, Liverpool 1917, Deputy Engineer Surveyor-in-Chief, London H.Q. 1921, Engineer Surveyor-in-Chief 1926, retired 1931. Vice-President, I.Mar.E. 1927.

I.S.O. 1928.

**Sir John Niven, Kt**

Born Scotland 7 January 1877. Died 20 January 1947.

Partner, Andrew Weir & Co, London (joined them in 1892): Chairman, Baltic Mercantile Exchange 1935-7: Member, Council of Shipping of the U.K.: Director, Commercial Services, Ministry of Shipping.

Knighted 1937.

**Major Sir Percival Reuben Reynolds, K.B.E.**

Born 1876. Died 28 November 1965.

Director, B.C.R. Factories Ltd, Welwyn Garden City 1937: 1914-9 European War, M.E.F. and E.E.F. (mentioned in despatches): President, National Association of British and Irish Millers 1925-7.

O.B.E. 1921: K.B.E. 1928.

**A. L. Moore (secretary).**

## SHIPBUILDING LOANS COMMITTEE (1939)

**Sir Alan Rae Smith, K.B.E. (Chairman)**
Born 28 May 1885. Died 11 July 1961.
Chartered Accountant, senior partner of Deloitte Plender Griffiths & Co: Director of Savoy Hotels Ltd: Financial Adviser to Ministry of Shipping 1939-41, and Ministry of War Transport (later Ministry of Transport) from 1941: Member of numerous official committees 1929-52. World War I, Major R.A.S.C. Retired 1956.
O.B.E. 1918: K.B.E. 1948: Kt 1935.

**Frank Charlton, F.C.A., F.I.C.S.**
Born 1890. Died 22 September 1965.
Articled to Newcastle Chartered Accountants: Joined Price, Waterhouse & Co 1913: White Star Line, Liverpool, 1921, as Chief Accountant, transferred to London 1928 as General Manager and Director: Secretary, Royal Mail Steam Packet Voting trustees 1930-3: Director, Cunard White Star Line and Cunard S.S. Co Ltd, 1934: Board of Trade and Minstry of War Transport, 1939-43: Director, Furness, Withy & Co Ltd, 1944, rising to Chairman [also Chairman, Economic Insurance Co Ltd and a director of many Furness Withy Group companies (Royal Mail Lines, Pacific S.N. Co, Prince Line, Shaw Savill & Albion Co, British Maritime Trust, etc)]: retired 1962

**Sir George Perrin Christopher, Kt** — *see above.*

**George Chester Duggan, C.B., O.B.E.**
Born 1887. Died 15 June 1969.
Entered Civil Service 1908: Admiralty 1908-10 and 1914-6: Ministry of Shipping 1917-9: Chief Secretary's Office, Dublin Castle 1910-4, 1919-21: Assistant Secretary, Ministry of Finance, Belfast 1922-5: Principal Asst Secretary, Ministry of Finance, Belfast 1925: on loan to Ministry of War Transport, London, 1939-45: Comptroller and Auditor-General for Northern Ireland, 1945-9.
O.B.E. 1918: C.B. 1930.

**Sir John Niven, Kt** — *see above.*

**Sir (Philip) Dennis Proctor, K.C.B.**
Born 1 September 1905.
Entered Civil Service 1929, serving in Treasury 1930-50. Re-entered Civil Service 1953, retired 1965 as Permanent Secretary, Ministry of Power. Managing Director, The Maersk Co Ltd, 1951-3: Director Williams Hudson Ltd, 1966-71.
C.B. 1946: K.C.B. 1959.

**C. T. Plumb (secretary).**

### Appendix e

## VESSELS SUBMITTED FOR SCRAPPING BUT DECLINED BY THE SHIPS REPLACEMENT COMMITTFF

**ANGELE MABRO** (ex PLM10, Easingwold), 3,154/98. 3 or 6.7.1940 sailed Bilbao for Cardiff, iron ore. 8.1.1941 posted missing.
**ANTHIPPI N. MICHALOS** (ex Cromerton, Novorossia), 3,298/05. 22.12.1942 sunk in collision, 53.10N 53W. Wabana for Belfast and Cardiff, iron ore.
**ANTONIOS MICHALOS** (ex Georgios K. Saliaris, Corby), 3,514/01. 1938 broken up.
**CESAR MABRO** (ex Rounton), 2,646/94. 1937 broken up.
**CLARA HINTZ** (ex Helmwige, Lizzie, Hardwick), 1,096/80. 1937 broken up.
**EUGENA CAMBANIS** (ex Nea Ellas, Sir W. T. Lewis), 3,470/98. 29.11.1940 in distress, abandoned, heavy weather in 46.53N 48.37W. St Anne des Montes and Sydney, NS, for Belfast. Presumed foundered 300 miles E of St John's, Nfld, no survivors.
**HELIOS** (ex Tremayne), 1,541/86. 1936 HELNY, 1942 FRIGG. 28.4.1944 struck wreck and sank off Kiel. Emden for Sundsvall, coke.
**HYDRAIOS** (ex Parnon, Isford, Lothian), 4,476/02. 6.7.1943 torpedoed by U198 in 24.44S 35.12E. Port Said and Aden for Lourenco Marques, rock salt ballast.
**ITHAKI** (ex Anne, Aenne Rickmers), 4,083/11. 1939 MOLDOVA, 1954 JAGRAHAT, 1955 MOLDOVA. 1958 broken up.
**ITHAKOS** (ex Ellaline), 3,916/06. 1942 SIGURD FAULBAUM, 1946 ITHAKOS. 1953 broken up.
**MAJA** (ex City of Montdidier, Folda, Soldier Prince), 1,283/83. 4.7.1936 stranded during fog, sank outside Trysunda. Hartlepool for Kopmenholmen, coal.
**MARECHAL FOCH** (ex Buda II), 2,853/05. 1936 HA-VEN, 1938 FOCH. 8.12.1941 seized at Yokohama, renamed HOSHI MARU. 25.7.1945 mined and sunk in Maizuru Bay.

**EUGENIA CAMBANIS.** *(R. J. Scott)*

**MARTIS** (ex William Balls), 2,483/94. 6.1940 scuttled as blockship, Scapa Flow.
**MICHALOS XILAS** (ex Greystoke Castle), 3,828/06. 1936 broken up.
**MIMOSA** (ex Michael, Kirnwood), 3,071/05. 1950 broken up.
**NAFTILOS** (ex Leamington, Tynemede, Blue Jacket), 3,531/04. 15.7.1940 sunk by gunfire from U34 in 48.5N 10.25W. San Nicolas for Dublin, grain.
**NJEGOS** (ex Norman Isles, Suruya), 4,376/08. 9.6.1944 scuttled as blockship, Arromanches.
**NORA** (ex Langdon), 1,252/82. 1939 MAGDALENA. 1952 broken up.
**OKEANIA** (ex Candidate, Crown of Castile, Ormiston), 4,843/07. 8.4.1940 mined and sunk in 51.16.8N 2.3.2E. Rotterdam for La Plata, ballast.
**PAGASITIKOS** (ex Sicily), 3,457/14. 1937 FRANCOIS, 1937 VIGO, 1939 CASTILLO DE ANDRADE, 1948 ANTARTICO. 6.10.1959 grounded entering Santander. Tenerife for Santander, manganese ore. T.L.
**PATRA** (ex Orion, Hillcrag), 3,256/96. 1937 STYLIANI. 6.4.1941 bombed and sunk at Piraeus.
**POLI** (ex Thraki, Johanna Blumberg, Crest, Dorotea, Crest), 2,861/97. 2.4.1937 sunk by gunfire from Nationalist cruiser BALEARES 28 miles off San Antioco during Spanish Civil War. Bagnoli for Oran and Melilla.
**VIDOVDAN** (ex Braunfels), 5,581/06. 15.12.1939 wrecked at Great Natuna, North Natuna Island, in 5N 108E. Miike for Sourabaya, ballast.

Note: By whom these vessels were submitted is not known.

## Appendix f

## REPORT OF THE COMMITTEE

The following is the full report of the Ships Replacement Committee, which was appointed by the Board of Trade in April, 1935, under the British Shipping (Assistance) Act, 1935:

Part II of the British Shipping (Assistance) Act, 1935, under the provisions of which the Ships Replacement Committee was constituted, expired on the 25th February, 1937, and the Committee submits the following report:

2. *The Scrap and Build Scheme* — The above enactment gave legislative sanction to what has usually been known as the "scrap and build" scheme. It empowered the Board of Trade, with the concurrence of the Treasury, and upon recommendations made by the Committee, to assist approved proposals for the building and modernisation of ships by making advances bearing interest at a rate not exceeding 3 per cent per annum and repayable within a period not exceeding 12 years. The maximum limit of advances was fixed at ten million pounds. The over-riding requirement was that no proposal might be approved unless the Board of Trade were satisfied that approval would "promote the general interests of British shipping", but the distinctive feature of the scheme, and the one from which it took its name, was the provision that two tons of shipping must be scrapped for every ton built with assistance granted under the Act, and that one ton must be scrapped for every ton modernised. Only vessels employed in the carriage of commercial cargoes were eligible for assistance, and even such vessels were excluded if they were employed mainly in the coasting trade or in maintaining regular services in the near Continental trade.

3. *Functions of the Committee* — From this brief description it will be seen that the main object of the scheme was the promotion of the general interests of British shipping by reducing redundant tonnage and by assisting British shipowners to improve and modernise their fleets, and this fact was further emphasised by the composition of the Committee. A subsidiary object, of course, was the revival of the shipbuilding industry and the lessening of unemployment, and it is a source of satisfaction to the Committee that many of the vessels built under the scheme were built in yards situated in depressed areas. The main function of the Committee was to submit the recommendations upon which alone the Board of Trade could make advances. They scrutinised the applications, and before making any recommendations, satisfied themselves that they complied with all the requirements of the Act, and that it would promote the general interests of British shipping if the proposals were approved.

4. *Finance Sub-Committee* — In order that their recommendations might cover all aspects of each case, the Committee undertook to examine each application in its financial aspect, and to record an opinion whether, having regard to the circumstances in which the scheme had been introduced, the applicant might reasonably be regarded as worthy of credit under the Act. In many cases this task involved the examination of unpublished balance-sheets and other confidential documents, and it was accordingly entrusted to a sub-committee consisting of the chairman and Sir Percival Reynolds, which reported the results, but not the details, of its investigations to the main Committee. The task of the sub-committee was greatly facilitated by the assistance of Mr. Gershom Davis, Board of Trade accountant, and the thanks of the Committee are due to this gentleman.

5. *Specification of new vessels* — It was an essential part of the scheme that shipowners taking advantage of the Act were free to settle the specifications for the work of building and to select the firm to carry out the work. But having regard to the objects of the scheme, the Committee required applicants in all cases to furnish outline specifications and estimates of costs of the vessels which it was proposed to build. These specifications were scrutinised by the technical experts on the Committee in order that the Committee might satisfy themselves in each case that the design was generally in line with modern developments, and that the proposed new ship was of modern economical type and conformed to a good standard of specification at a reasonable cost. As has been explained, the Committee had no power to stipulate for any particular type of propelling machinery, but it is of interest to note that 55 per cent of the tonnage constructed under the scheme is steam and 45 per cent Diesel.

6. At the request of the Board of Trade, the Committee undertook to draw the attention of successful applicants to the necessity of providing in their new ships *(a)* modern, efficient steering gear and *(b)* good crew accommodation. This was done in all cases, and the Committee are glad to be able to report that all the shipowners concerned readily complied with requirements in these matters.

7. *Work of the Committee* — The Committee held 26 meetings at which they considered 74 applications for assistance under the scheme. These applications covered proposals for the building of 95 vessels, aggregating 334,794 tons gross, at an estimated cost of £6,463,350, of which the applicants desired to borrow £6,228,353.

8. *Modernisation* — No applications were received covering the modernisation of existing vessels. Doubtless the main difficulty in the case of modernisation was that the loss on scrapping, which is dealt with in the next paragraph, made any proposals for modernisation uneconomical. There was also the further difficulty that the Act required a first mortgage on any ship on which an advance was made, and, as it is likely that the majority of ships in possession of Companies needing assistance were probably already subject to prior mortgage, the owners were unable to give the statutory security required.

9. *The problem of scrap tonnage* — The scarcity of scrap tonnage has throughout been the principal difficulty in the way of the scheme. Few applicants were able to offer tonnage of their own for demolition. Indeed, only six of the 97 vessels finally nominated for demolition were the property of the applicant Companies. The great majority of the applicants had to go into the open market for the scrap tonnage which they required. Suitable tonnage was never plentiful, and applicants always had to pay a premium over the scrap value, this premium becoming increasingly heavy as time went on.

10. *Place of demolition* — The existence of this premium at once gave rise to difficulties. It was the intention of the Act that vessels demolished under the scheme should ordinarily be broken up in this country, and the Committee began by insisting on this being done. But it soon became evident that if they continued this policy, it would add so much to the loss already facing applicants in the matter of scrap tonnage that the whole scheme would be in danger of breaking down. They recommended therefore that the Board of Trade should exercise the discretionary power conferred on them by section 3 (1) *(b)* (i) of the Act and allow up to 50 per cent of each applicant's quota of tonnage for demolition to be sold for breaking up abroad. This recommendation was accepted, but later on it became apparent that even this concession did not go far enough, and eventually it was found necessary to allow an entirely unrestricted market for the disposal of scrap tonnage.

11. *Non-competitive scrap tonnage* — The scarcity and rising price of suitable scrap tonnage for demolition brought another difficulty in its train. In the early months of 1936 it became evident that vessels which could not be said to have any real competitive value were being nominated for demolition in increasing numbers. The Act was silent on this matter, but it was clearly not in accordance with the spirit and intention of the Act that the Committee should accept for demolition tonnage which in their opinion was already out of commission, nor was it in the general interests of British shipping that the owners of this ancient tonnage (who were mostly foreigners) should be paid a heavy premium over the scrap value of their vessels wherewith to buy more up-to-date and competitive tonnage. The matter was brought to the

notice of the Board of Trade, and with their approval the Committee freely exercised their discretion in refusing to accept for demolition tonnage which in their opinion was too old to be really competitive.

12. With the improvement in freight markets which took place in the autumn of 1936, the scrap position became practically impossible. A few applicants completed their proposals by acquiring vessels which were actually trading and obtaining the Board's permission to continue trading until such time as the corresponding new vessels were completed, but as will be seen from the statistics given in paragraph 13, many proposals were ultimately abandoned owing to this difficulty.

13. *Applications recommended by the Committee* — As stated in paragraph 7, the Committee considered 74 applications for assistance received from 53 separate managing or owning Companies. Of these, 19 applications, covering 24 vessels of 86,139 tons gross, were rejected by the Committee, and six applications, covering six vessels of 15,762 tons gross, were withdrawn by the applicants concerned after preliminary consideration by the Committee. The remaining 49 applications, covering the construction of 65 vessels aggregating 231,893 tons gross, were recommended by the Committee, but 12 of these, covering 13 vessels of 44,486 tons gross, and part of another application, covering two vessels of 3,800 tons gross, were, after approval by the Committee, subsequently withdrawn or remained in abeyance on the 25th February. The main reason for these latter withdrawals was the acute difficulty in the later stages of the scheme of obtaining suitable tonnage for demolition to which reference has been made in the preceding paragraph.

14. *Effective results of the Scheme* — The net result is that 37 applications from 22 separate concerns were successfully completed. They cover the construction of 50 vessels of approximately 186,000 tons gross (339,000 d.w.). The total estimated cost of these 50 vessels was £3,664,360, and the total advances recommended by the Committee and approved by the Board of Trade and sanctioned by the Treasury, in respect of the construction of the vessels was £3,548,124 7s 9d. In connection with the proposals finally sanctioned, 97 vessels of 386,625 tons gross were nominated for demolition. Fuller details of this tonnage are given in paragraph 17.

15. *New tonnage* — Of the 50 vessels built or to be built with assistance under the scheme, 27 are steamers aggregating approximately 117,000 tons gross made up as follows:

19 tramp vessels — 4,300-5,800 gross tons each.
1 tramp vessel — 2,700 gross tons.
2 cargo liners — 2,000 gross tons each.
and 5 tramp vessels — 1,175-1,550 gross tons each, built specifically for the Baltic timber trade.

The remaining 23 vessels are motor vessels, 16 being ocean-going tramps of about 5,000 tons gross each, and seven small tramps of the coasting type, intended for near Continental trade. At the date of the termination of the Scheme (25th February, 1937), 38 of these vessels were completed and in service.

16. The ports at which the "scrap and build" vessels have been or are being constructed are as follows:

| | | Tons gross |
|---|---|---|
| *North-East Coast* | | |
| Sunderland ... ... | 24 vessels of approx ... | 98,000 |
| Hebburn-on-Tyne ... ... | 3 vessels of approx ... | 16,000 |
| Wallsend-on-Tyne ... ... | 3 vessels of approx ... | 15,000 |
| Haverton Hill-on-Tees ... ... | 3 vessels of approx ... | 12,000 |
| West Hartlepool ... ... | 1 vessel of approx ... | 4,000 |
| *Clyde* | | |
| Glasgow ... ... | 4 vessels of approx ... | 20,000 |
| Port Glasgow ... ... | 3 vessels of approx ... | 12,000 |
| *Rest of Scotland* | | |
| Burntisland ... ... | 2 vessels of approx ... | 7,000 |

The seven small motor vessels were built at Goole (4), Bristol (2) and Aberdeen (1).

17. *Tonnage demolished* — The 97 vessels nominated for demolition consisted of:

| | | Tons gross |
|---|---|---|
| 49 British (U.K. registered) ... ... | totalling ... ... | 238,467 |
| 48 foreign ... ... ... | totalling ... ... | 148,158 |

The foreign tonnage consisted of vessels of the following nationalities:

| | Tons gross | | Tons gross |
|---|---|---|---|
| 15 Greek ... ... | 55,156 | 1 Panamanian ... ... | 4,078 |
| 6 Finnish ... ... | 22,906 | 3 German ... ... | 3,862 |
| 5 Belgian ... ... | 13,934 | 1 Egyptian ... ... | 3,773 |
| 3 French ... ... | 9,858 | 2 Danish ... ... | 3,590 |
| 4 Swedish ... ... | 8,328 | 1 Estonian ... ... | 1,041 |
| 1 U.S.A. ... ... | 6,098 | 1 Danzig ... ... | 925 |
| 2 Latvian ... ... | 5,370 | | |
| 2 Norwegian ... ... | 4,944 | | 148,158 |
| 1 Yugo-Slav ... ... | 4,295 | | |

18. At the date of this report 94 vessels, totalling 374,707 tons gross, have either been demolished or have been disposed of to shipbreakers. 55 vessels, totalling 204,102 tons gross, have gone to shipbreakers in the United Kingdom, and 39, totalling 170,605 tons gross, have been sold to shipbreakers abroad. Of the latter, the main sales were to the following countries:

| | Tons gross |
|---|---|
| Japan — 9 vessels ... ... ... ... ... ... | 57,997 |
| Italy — 8 vessels ... ... ... ... ... ... | 46,426 |
| Holland — 6 vessels ... ... ... ... ... ... | 19,562 |
| Germany — 5 vessels ... ... ... ... ... ... | 14,040 |
| Danzig — 3 vessels ... ... ... ... ... ... | 12,862 |

Other sales were to shipbreakers in Sweden, Belgium, Denmark, Irish Free State, Latvia and Norway.

19. Arrangements for the disposal of the remaining three vessels have not yet been made by the shipowners concerned. These vessels, together with a few others which have been sold to breakers with forward delivery dates, are being permitted to trade until the corresponding new vessels are completed.

20. *General Summary* — More advantage would no doubt have been taken of the scheme had it been introduced a year or two earlier. As it was, the increasing scarcity and cost of scrap tonnage militated against the complete success of the scheme. Nevertheless, the Committee believe that it has fulfilled a useful purpose. The success of the scheme should not be judged merely by the fact that with the assistance afforded by the Act shipowners have been able to build 50 new ships. The new ships are ships of the most modern type, speedier and more economical to run than most of the ships now in commission. To this extent the scheme has contributed to the modernisation of the British mercantile marine. Moreover, for every new ton that has been built under the scheme two tons have been taken off the water, and of this demolished tonnage nearly 150,000 tons were foreign. Further, the orders for the new ships came at an opportune moment. They came at a time when shipbuilding was at a very low ebb. They were specially valuable to shipbuilders for that reason — particularly to those specialising in the building of tramps.

21. The Committee wish to place on record their deep regret at the death of their colleague, Mr A. E. Laslett. Mr Laslett's great knowledge derived from his long service as an engineer surveyor to the Mercantile Marine Department of the Board of Trade made him a most useful member of the Committee, and by his personality he had endeared himself to all his colleagues.

22. The Committee also desire to draw the attention of the Board of Trade to the excellent service rendered by the secretary, Mr A. L. Moore, and his assistant, Mr L. G. Child. Mr Moore's knowledge, ability and tact were quite invaluable to the Committee, and Mr Child was in every way a very capable understudy to Mr Moore.

(Signed) C. A. Innes
Francis G. Bryant
G. P. Christopher
John Niven
P. R. Reynolds

A. L. Moore, Secretary
15th March, 1937

## Appendix g

## INDEX OF SHIPBUILDERS

(including ships names and yard numbers)

Scrap & Build ships in CAPITALS, Shipping Loan in lower case letters.

**S.P. Austin & Son Ltd, Sunderland**

**Barclay, Curle & Co Ltd, Glasgow**
650 QUEEN ADELAIDE
659 QUEEN VICTORIA
660 QUEEN ANNE
662 DUNKELD

**Bartram & Sons Ltd, Sunderland**
272 NAILSEA COURT
274 NAILSEA MEADOW
276 NAILSEA MOOR
277 NAILSEA MANOR
284 Richmond Hill
285 Pentridge Hill

**Blythswood Shipbuilding Co Ltd, Glasgow**
60 Welsh Prince

**Burntisland Shipbuilding Co Ltd, Burntisland**
204 USKSIDE
213 DERRYMORE
239 Dan-y-Bryn
241 Elmdale
242 Ger-y-Bryn
243 Edencrag
245 Channel Queen
246 Coral Queen
247 Tudor Queen

**Caledon Shipbuilding & Engineering Co Ltd, Dundee**
384 Tottenham
385 Twickenham
386 Teddington

**William Doxford & Sons Ltd, Sunderland**
618 RUGELEY
620 RILEY
621 ROTHLEY
622 RIDLEY
623 RIPLEY
629 QUEEN MAUD
652 La Estancia
654 Rodsley
655 La Cordillera
659 Putney Hill
660 Tower Grange
661 Rawnsley
663 Rookley
665 Duke of Athens
666 Reaveley
668 Antar

**Furness Shipbuilding Co Ltd, Haverton Hill-on-Tees**
247 BRADFORD CITY
249 CORNISH CITY
262 SYRIAN PRINCE

**Goole Shipbuilding & Repairing Co Ltd, Goole**
312 ASHANTI
313 BENGUELA
314 CABENDA
315 LOANDA

**William Gray & Co Ltd, Hartlepool**
1061 CRESSDENE
1095 Elmdene
1100 Industria
1104 Duke of Sparta

**ELMDENE** (see page 49). *(L. Dunn)*

**William Hamilton & Co Ltd, Port Glasgow**
425 ARABIAN PRINCE
427 DARLENY
429 DARLYON
440 Lulworth Hill
441 Kingston Hill

**Hawthorn, Leslie & Co Ltd, Newcastle**
601 WINDSORWOOD
602 YORKWOOD
609 BALMORALWOOD

**Charles Hill & Sons Ltd, Bristol**
244 PURBECK
251 CASTLE COMBE

**Sir James Laing & Sons Ltd, Sunderland**
711 LOWTHER CASTLE
713 LANCASTER CASTLE
727 Beechwood

**John Lewis & Sons Ltd, Aberdeen**
141 CROMARTY FIRTH
145 Worthtown

**Lithgows Ltd, Port Glasgow**
921 Broompark
922 Glenpark
925 Ribera
929 Temple Arch
931 Temple Inn
937 Rembrandt

**Lytham Shipbuilding & Engineering Co Ltd, Lytham**

**William Pickersgill & Sons Ltd, Sunderland**
232 HYLTON
241 Daydawn
243 Stanmore

**J. Readhead & Sons Ltd, South Shields**
519 North Britain

**Henry Robb Ltd, Leith**

**Short Brothers Ltd, Sunderland**
445 SPRINGWEAR
446 SPRINGWOOD
447 SPRINGWAVE
450 BIDDLESTONE
451 SPRINGTIDE
452 SPRINGDALE
463 Newbrough

**Smith's Dock Co Ltd, Middlesbrough**

1066 Norman Prince
1067 Lancastrian Prince
1068 Tudor Prince
1069 Stuart Prince

**Swan, Hunter & Wigham Richardson Ltd, Wallsend-on-Tyne**

1513 HOPESTAR
1533 HOPECASTLE
1535 HOPECROWN
1638 Hopetarn

**Joseph L. Thompson & Sons Ltd, Sunderland**

573 ST HELENA
574 ST MARGARET
575 ST CLEARS
576 STARCROSS
582 ST ROSARIO
598 Graiglas
600 St Essylt

## INDEX OF ENGINE BUILDERS

Aitchison, Blair Ltd, Greenock
Atlas Diesel A/B, Stockholm
Barclay, Curle & Co Ltd, Glasgow
British Auxiliaries Ltd, Glasgow (later British Polar Engines Ltd)
Central Marine Engine Works (Wm Gray & Co Ltd), West Hartlepool — CMEW.
G. Clark (1938) Ltd, Sunderland
Crossley Brothers Ltd, Manchester
William Doxford & Sons Ltd, Sunderland
Humboldt-Deutzmotoren A.G., Koln-Deutz
J.G. Kincaid & Co Ltd, Greenock
Mirrlees, Bickerton & Day Ltd, Stockport
North Eastern Marine Engineering Co Ltd, Newcastle & Sunderland — NEMEC.
(later North Eastern Marine Engineering Co (1938) Ltd)
Nydqvist & Holm A/B, Trollhattan
Parsons Marine Steam Turbine Co Ltd, Wallsend-on-Tyne
Rankin & Blackmore Ltd, Greenock
Richardsons, Westgarth & Co Ltd, Hartlepool & Middlesbrough
D. Rowan & Co Ltd, Glasgow
Ruston & Hornsby Ltd, Lincoln
Smith's Dock Co Ltd, Middlesbrough
Swan, Hunter & Wigham Richardson Ltd, Wallsend-on-Tyne
Joseph L. Thompson & Sons Ltd, Sunderland
White's Marine Engineering Co Ltd, Newcastle

**Appendix h**

## MACHINERY

Whilst many vessels were fitted with straightforward triple expansion or diesel machinery, quite a few employed more advanced steam machinery to achieve economies.

As steam pressure falls it loses its heat content and gets progressively wetter. With this goes loss of efficiency. As a first stage superheating to 600°F gave a 15% economy when compared with saturated steam.

Technical developments had produced the "Woolf" double compound principle, examples being the Lentz, Fredriksstad 'Steam Motor' and Christiansen & Meyer. Low pressure turbines included the Bauer-Wach with flexible hydraulic coupling, the Elsinore and Parsons with spring coupling and Gotaverken's turbo-compressor. Intermediate superheat was featured in the Reheater.

The following machinery featured in the Scrap and Build programme:

*Reheater* — a design by the North Eastern Marine Engineering Co (NEMEC). The first newbuilding was the LOWTHER CASTLE, previous installations having been conversions. Steam at 220 lb/sq in and 750°F from the boiler passed through a heat exchanger (the 'Reheater') before entering the HP cylinder at 600°F. On leaving the HP cylinder it passed through the other side of the reheater, the temperature being

raised by 140-200°F before going to the IP and LP cylinders. This kept the steam dry and gave a 10% economy over normal superheating. Exhaust steam leaving the LP cylinder still retained some superheat, hence eliminating cylinder condensation which was the greatest source of thermal inefficiency in reciprocating engines.
*Gotaverken turbo-compressor* — after leaving the LP cylinder, and before entering the condenser, steam drove a turbine coupled to a compressor. This compressor raised the steam pressure between the HP and IP cylinders.
*White compound engine* — a twin (two HP and two LP cylinders) high speed compound engine driving through single reduction gears, exhausting to a low pressure turbine driving through double reduction gears, all mounted as a single unit. The design included reheat at each stage, between HP and LP, also between LP and turbine. The first installation was in the ADDERSTONE (ex Boswell), and the BIDDLESTONE was the first in which the principal auxiliary machinery was driven off the main engine. An 8,000 dwt vessel, she could achieve 10 kts on 12 tons of coal a day.
*Parsons Simplex turbine* — a new design, the first example being exhibited at Glasgow's Empire Exhibition and then fitted in the HOPESTAR. Two forward turbines (HP and LP) gave 2,000 shp, the design being intended for cargo vessels, and drove a propeller at 75 rpm through flexible couplings. The turbines, including two reverse turbines, and condenser were mounted on the gearcase as a compact unit weighing 56 tons. Superheated steam was taken at 285 lb/sq in and 750°F.

## Appendix i

## FAIRPLAY "STANDARD" SHIP

In reviewing changes in the prices of secondhand ships 'Fairplay' included details of newbuilding costs. For this a theoretical ship was specified, a single deck three island vessel built to a very plain specification (no Grain Act requirements, deep tank or donkey boiler), three boilers and triple expansion engines giving 9 kts on 25 tons of coal per day. Dimensions were 380 ft × 49 ft × 29 ft and she loaded 7,500 dwt on a draft of 24 ft 6 ins.

Estimated newbuilding costs were:

| | June | December | | June | December |
|---|---|---|---|---|---|
| 1914 | | £ 54,375 | 1931 | £ 66,750 | £ 63,500 |
| 1915 | | 110,625 | 1932 | 63,250 | 62,250 |
| 1916 | | 172,500 | 1933 | 62,250 | 63,250 |
| 1917 | 195,000 | | 1934 | 63,750 | 64,500 |
| 1920 | | 225,000 | 1935 | 65,750 | 71,600 |
| 1921 | 135,000 | 97,500 | 1936 | 72,000 | 78,000 |
| 1922 | 75,000 | 67,500 | 1937 | 100,000 | 108,000 |
| 1923 | 73,250 | 72,000 | 1938 | 108,000 | 95,000 |
| 1924 | 75,000 | 68,000 | 1939 | 100,000 | 119,000 |
| 1925 | 65,000 | 60,000 | 1940 | 132,000 | 156,000 |
| 1926 | 60,000 | 64,500 | 1941 | 156,000 | 156,000 |
| 1927 | 66,000 | | 1942 | 156,000 | 156,000 |
| 1928 | 65,250 | 66,500 | 1943 | 162,500 | 162,500 |
| 1929 | 67,250 | 67,750 | 1944 | 165,600 | 165,600 |
| 1930 | 67,750 | 67,750 | 1945 | 180,000 | 180,000 |

The effect of the Depression on prices was that such a new, ready vessel would fetch about £32,000 on the open market in 1932-3 compared with the newbuilding cost of about £63,000. In January 1931 when the cost of such a vessel built to a plain specification was £67,750 an even barer specification was priced at £48,750 and to a very good specification £80,250.

By the 1930s the specification was out-of-date, technical developments had reduced fuel consumption and the shelter decker predominated in the Scrap & Build and Shipping Loan programmes. But it was to be 1945 before a new 'Fairplay' standard ship reflected the developments of the 1930s and typified the vessel which featured to a large degree in both the Scrap & Build and Shipping Loan schemes, also their close relatives the Liberty and Victory ships. An open shelter deck design loading 9,500 dwt on a draft of 25 ft 8 ins her dimensions were 425 ft × 58 ft × 29 ft (38 ft to shelter deck). Tonnages were 5,000 gross and 3,000 nett whilst a diesel of 3,300 bhp gave 12.5 to 13 kts on 13 tons of fuel daily.

The modern equivalent of this ship would probably be Austin & Pickersgill's SD9, small sister to the SD14, a tween deck design for which no orders have been placed. Her designed deadweight is 9,000 tons on 23 ft 2 ins draft, dimensions of 387 ft × 61 ft × 21 ft 4 ins (32 ft 9 ins to upper deck) and a diesel of 5,120 bhp gives 14.2 kts. Her price today (1982) is about £5.5 million.

In commenting on the influence on prices from various extras and design features Mr A.L. Ayre of the Burntisland Shipbuilding Co Ltd wrote (quoted from "British Shipping Finance"):

It may be a convenient way, and in fact quite sound, to buy coal at so much a ton, just as eggs can suitably be rated at so much per dozen, but to value ships for comparative purposes, one with another, at so much per deadweight ton, is a positive absurdity that has crept into the business of buying and selling ships during the past few years. The use of expression is perhaps more general in this country than it is abroad; at least, one might judge this from experience of negotiations with both home and foreign buyers. It can never be a sound method of comparing *values* of ships, however much it may be used by the lesser experienced to compare prices. The continued use of the expression can only lead the ship designer in one direction — that is, to design vessels that will cost the smallest amount per ton deadweight, regardless of anything else, and particularly disregarding efficiency and economy in service.

It is surprising to find in certain directions, where it is a business or profession of people to determine values of ships, a lack of knowledge as to how the absurd term, "price per ton," varies with the size of the ship, let alone any of the other factors. Careful estimates made for a series of ships throughout the range of deadweight up to 8,000 tons for single-deck ships, whose drafts and speeds are in uniformly increasing ratio, and the specification throughout of the same class, being varied in quantities only, to meet suitably the increasing size of ship, show variations of quite important amounts, especially as the vessels reduce in size. For instance, if an 8,000 tonner is taken at a certain rate per ton, it is found that for a similar class vessel, but reduced deadweight, at 4,000 tons, the rate per ton has increased by £1. Another addition of £1 is experienced at 3,500 deadweight and further additions of £1 each time are found at deadweights of 2,750, 2,250, and 1,900, while below this the rates per ton increase at a very rapid rate. This comparison of cost per ton on size is, however, the simplest of all, and in this illustration the various important factors which control the cost of ships have been taken most carefully into strict relative account, so that the variation is as true as it is possible to make it.

But once we leave this simple strict relative comparison and let the subject become complicated with differences in type, specification, draft, and speed, then immediately it becomes impossible to use the price per ton as a rational or sensible means of judging the value of a ship or of comparing the value of one ship with another. A lower deck is worth, roughly, 10s. per deadweight ton in average size cargo vessels. This item has an effect in two directions — it reduces the deadweight by the amount of its own weight and increases the cost.

Again, take specification: the complete equipment of a vessel for Grain Act purposes is worth about 3s to 5s per deadweight ton according to size. In a 2,500-tonner the difference between single winches and derricks and double winches and double derricks amounts to nearly 10s per ton deadweight, and a small donkey boiler alone would be worth a further 5s per ton. Draft is another important factor, and one which is often disregarded when comparing prices for values of vessels which may otherwise be similar as to size, etc. A vessel to carry 6,000 tons deadweight on 22 ft 6 in draft could have length and breadth of 339 ft by 48 ft, but although retaining all other factors, such as type, specification, speed, etc, if it is desired to carry the same deadweight on only 21 ft 6 in draft, it will be necessary to increase the length and breadth to 348 ft by 49 ft 3 in. The depth would, of course, be reduced by about 1 ft 2 in, but depth, it must be remembered, is the cheapest dimension of the three, length being the most expensive. In fact, the vessel of 21 ft 6 in draft requires about 50 tons of steel in excess of the amount required to build the vessel of same deadweight on 22 ft 6 in. There are very many other items altered as a result of the increased dimensions — for instance, timber, and possibly the equipment — while the labour cost will also be greater. The value of one foot less draft in a vessel of this size would be about 3s to 4s per deadweight ton.

Speed is one of the most important and expensive of the factors which cause variation in the cost per ton deadweight. To take the 6,000 ton vessel again as an example and compare the changes which are necessary for increasing the speed from 10 knots to 11 knots on load trial, the deadweight, draft, and specification being identical in both cases, we have the following important alterations to provide for:-

I.—A finer block co-efficient;

II.—An increase of about 39 per cent in indicated horse power, requiring larger engines and boilers, the increased weight of which will amount to nearly 100 tons;

III.—Because of the finer block co-efficient and the increased weight of machinery, the dimensions will require to be increased so as to maintain the deadweight;

IV.—The increase of dimensions will have similar effect on the cost as was mentioned in previous illustration; in fact, in this instance, the extra weight of steel required is about 60 tons.

The net result is to increase the cost of the hull by about 4s per deadweight ton, and the machinery by about £3,600, or 12s per deadweight ton, or 16s per ton all told.

Again, comparing two vessels of identical deadweight, although of similar type and specification, but one having a draft one foot shallower, also one knot additional speed than the other, the difference in value is practically 20s per deadweight ton.

Difference in block co-efficient may make a substantial alteration in the cost of a ship; for instance, to reduce the block co-efficient in a 6,000 deadweight vessel by 0.02 will cause a loss of about 200 tons of deadweight, and therefore increase the cost by about 6s per ton.

There are various other factors which could be referred to as affecting the rate-per-ton method of valuing ships — in the aggregate they are of very substantial magnitude — but sufficient examples have, perhaps, already been given to show how ridiculous it is to make comparisons on such an irrational basis."

## Appendix j

## MERCHANT SHIP RESERVE

1. Ships purchased:
The following four were the only vessels acquired for the Reserve. They were taken over by the Admiralty, later being managed by the firms shown in brackets after their names on behalf of the M.o.W.T.

*MANCHESTER PRODUCER* (ex Start Point), 5106/16. Manchester Liners Ltd, Manchester (£17,500). Renamed *BOTWEY* (mgrs: P. Henderson & Co, Glasgow). 26.7.1941: torpedoed by U141 in 55.42N 9.53W, convoy OS1. Ellesmere Port for Port Sulphur, ballast.

*PENTRIDGE HILL* (ex African Prince, Glennevis), 5119/17. Dorset S.S. Co Ltd (Counties Ship Management Co Ltd), London (£26,000). Renamed *BOTLEA* (mgrs: Sir William Reardon Smith & Sons Ltd, Cardiff). 9.1939: Q-ship H.M.S. LAMBRIDGE. 2.1941: Armed Merchant Cruiser. 12.1941: returned to M.o.W.T. 30.12.1945: scuttled in 55.30N 11W with ammunition (gas shells).

*TRADE* (ex Eurypylus, Indrakuala), 5848/12. Continental Transit Co Ltd, London (£20,000). Renamed *BOTAVON* (mgrs: Sir William Reardon Smith & Sons Ltd, Cardiff). 2.5.1942: torpedoed by aircraft of I/KG26 off Murmansk, 73.2N 19.46E, convoy PQ15. 3.5.1942: sunk by escort. Middlesbrough for Murmansk, Government stores. 73 crew, 21 lost.

*TRANSIT* (ex Molton), 3091/19. Continental Transit Co Ltd, London (£21,500). Renamed *BOTUSK* (mgrs: Sir William Reardon Smith & Sons Ltd, Cardiff). 31.1.1941: Mined 6 miles NE of Rona Isl. St John, NB, for Cardiff, grain. 38 crew, 4 lost.

**BOTLEA** as H.M.S. LAMBRIDGE. *(National Maritime Museum)*

2. Ships under negotiation:
The following ships were reported sold to the Government for the Reserve but the sales either fell through or were aborted by the outbreak of war which made the scheme obsolete.
*CELTIC STAR* (ex Celticstar, Camana), 5575/18. Union Cold Storage Co Ltd (Blue Star Line Ltd), London (£50,000). 29.3.1943: torpedoed by Italian submarine FINZI in 4.16N 17.44W. Manchester for Montevideo & Buenos Aires, general. 64 crew & 2 passengers, 2 lost & 1 p.o.w.
*EGBA*, 4989/14. Elder Dempster Lines Ltd, Liverpool. 1943: sold to M.o.W.T. (mgrs: Elder, Dempster Lines Ltd, Liverpool), renamed *EMPIRE SEVERN.* 12.10.1946: scuttled in 58.18.30N 9.37E with ammunition.
*POLYCARP*, 3577/18. Booth S.S. Co Ltd, Liverpool. 2.6.1940: torpedoed by U101 in 49.19N 5.35W. Para for Liverpool, general.
*REDSTONE* (ex Margari, Orbe, Wye Crag, War Crag), 3110/18. Phoenix Sg Co Ltd, London (£22,500). 2.5.1940: scuttled as blockship at Scapa Flow. 9.1948: raised and broken up by Arnott, Young Ltd, Cairnryan.
*STANGRANT* (ex Clan Grant, Cambrian Marchioness, Port Macquarie), 5804/12. Stanhope S.S. Co Ltd (J.A. Billmeir & Co Ltd), London (£20,000). 13.10.1940: torpedoed by U37 in 58.27N 12.36W. Hampton Roads for Belfast, steel & scrap. 38 crew, 8 lost.
*TRECARREL* (launched as War Lilac), 5271/19. Hain S.S. Co Ltd, London (£30,000). 4.6.1941: torpedoed by U101 in 47.10N 31W. Hull for Father Point, ballast. 43 crew, 4 lost.
*TREFUSIS* (ex War Aconite), 5299/18. Hain S.S. Co Ltd, London (£30,000). 5.3.1943: torpedoed by U130 in 43.50N 14.46W, convoy XK2. Pepel for London, iron ore. 47 crew, 3 lost.
*TREKIEVE* (launched as War Mallard), 5244/19. Hain S.S. Co Ltd, London (£30,000). 4.11.1942: torpedoed by U178 in 25.46S 33.48E. Bombay for Durban/U.K., manganese ore & general. 48 crew, 3 lost.

# INDEX OF COMPANIES

(shown as sub-headings)

**USKSIDE** in typical wartime condition (see page 32). *(National Maritime Museum)*

# INDEX OF SHIPS